A technical history of the Río Tinto mines: some notes on exploitation from pre-Phoenician times to the 1950s

A technical history of the Río Tinto mines: some notes on exploitation from pre-Phoenician times to the 1950s

Leonard Unthank Salkield

Edited by Maurice J. Cahalan

The Institution of
Mining and Metallurgy

Printed in England by Barnes Design + Print Group, Maidenhead

Published at the office of
imm | The Institution of Mining and Metallurgy
44 Portland Place London W1N 4BR England

ISBN-13: 978-94-010-8017-0 e-ISBN-13: 978-94-009-3377-4
DOI: 10.1007/978-94-009-3377-4

Contents

Editor's note

Leonard Salkield was born in Hoddesdon, Hertfordshire, in 1901 and graduated from the Royal School of Mines, London, in 1922. He worked at Enthoven's lead refinery in Rotherhithe and Eastern Smelting Company's tin smelter in Penang (Malaysia) before joining the Rio Tinto Company in 1928 to work at its copper smelter at Port Talbot in Wales. He transferred to the Company's Ewell laboratory in 1929 and to the Rio Tinto Mines in 1930. He returned to England in 1937 during the Spanish Civil War and, after working on a wartime magnesium metal project, returned to Rio Tinto Mines in 1946 to spend the remainder of his professional career there.

He became interested in the history of activities at the site soon after first going to Rio Tinto and his work on the development of metallurgical processes was a natural complement to his enquiry into the nature of the earlier operations, which had left so much evidence of smelting on a very large scale. In 1957 he began to assemble notes for Eric Linklater, who had been invited to write a company history. That project did not proceed and when David Avery was commissioned to write a history of the Company* it was natural that the then Chairman of Rio Tinto-Zinc, Sir Val Duncan, should ask Leonard to assist in providing insight into the prehistory of the mines.

This involvement stimulated him to further efforts in tracking down information relating to the 4000 or more years of activities. Work by professional archaeologists on copper smelting had started at Rio Tinto around this time, following earlier dramatic successes in unravelling the ancient technology in the Middle East, and Leonard collaborated enthusiastically with the various groups that become involved — his unrivalled knowledge of the changes made by the operations of the past century was of considerable assistance to them.

In 1962 he received the O.B.E. He continued as a consultant to Rio Tinto Patiño, one of the successor companies to The Rio Tinto Company, for many years after 'retirement', living in Spain near Gibraltar and pursuing the collection of information relative to activities at the Rio Tinto mines. When he finally retired to live in England near his family he embarked on the task of preparing his assembled material for publication. With his keen desire to resolve all possible ambiguities, access to the British Museum and other sources of additional information delayed him in this task and, sadly, it was not completed before his death in 1985.

* Avery D. *Not on Queen Victoria's birthday: the story of the Rio Tinto mines* (London: Collins, 1974), 464 p.

To us has fallen the stimulating and challenging opportunity of editing and presenting the information that he collected in a form of which we hope that he would have approved. It was never his aim to duplicate the history of The Rio Tinto Company activities so ably presented in David Avery's book, but he felt that a volume recording important technical information would be of value and we agree wholeheartedly with him on that. This is not intended as a history in the accepted sense: rather, it is a volume of technical notes set in a chronological order, roughly to the end of the Rio Tinto Company operations. Even so, there is a considerable amount of reference forward and back, especially since places and areas are frequently identified by names that they assumed in recent times.

Whereas David Avery drew mainly on records from the old London office of the Rio Tinto Company in writing his history of the Company, Leonard Salkield searched through records at the mines and differing perspectives are given on some episodes. Clerical staff at the mines kept meticulous operating records, and although it must be said that many of the numbers contain far fewer significant figures than decimal points suggest, such records have provided a wealth of interesting detail for those technical notes.

The great interest of Rio Tinto to industrial archaeologists has been in unravelling the metallurgy employed there and, even in comparatively recent times, metallurgical process development and operation has been of special interest because of the unusual complexity of the ores and the novelty of processes to treat them. With his unrivalled insight into metallurgy at Rio Tinto, Leonard naturally emphasized metallurgy in his writings, but he also assembled interesting information on mining and infrastructure.

We have edited the notes to provide a general account, albeit slanted towards metallurgical operations, supplemented with more detailed appendices. There are instances where Spanish words or phrases are not translated because no English words convey quite the same meaning, or the Spanish is widely used and understood in the mineral industry, or for both reasons. For place names for which there is a widely used anglicized version, such as Seville for Sevilla, we have used the anglicized version. Rio Tinto, literally 'red river', was used as the name of quite separate townships in the mining area, in referring to the general area of the mines, and sometimes as an abbreviation for company names. Leonard Salkield always referred to the stream itself as 'the Tinto river' and we have also used that convention; other references to Rio Tinto should be clear from their context.

Many of the references listed are not readily available to the general reader. An exception is David Avery's *Not on Queen Victoria's birthday*, which covers much of the subject matter of these notes for the lay reader and provides an admirable account of social, political and economic issues concerning operation of the Rio Tinto mines from the sixteenth century. We have not been able to check some of the original documents and Leonard Salkield's translations and deductions from such sources are given verbatim. We have established that Rua Figueroa* is a major source for the detailed information and statistics given in Chapters 3–5.

Some of the terminology used at Rio Tinto may need a little explanation. In mining, levels are invariably referred to as 'floors': whether this derives from the Spanish 'piso' or was introduced from United Kingdom coal mining is not clear. The massive pyrites, as mined, is referred to as 'mineral', 'ore' or 'pyrites'; if of significant copper content, it is called cupreous mineral or 'cupreous'; low copper pyrites is thus non-cupreous. Country rock enclosing the lodes is porphyry; in some areas the copper content means that this is copper ore. Heap leaching of copper from massive pyrites is known as 'washing', hence 'washed ore' is pyrites recovered from the leaching heaps.

In a preface to his draft Leonard drew attention to the help that he had received in compiling his notes from the late Professor David Williams on geology, Mr Stanley Tong on the Atalaya opencast, Dr P.T. Craddock on archaeology, the late Col. R.F. Lethbridge, Mr John Hunt, Sr D. Eduardo Figueroa Poyatos and Mr Ivor Herbert. In preparing the material that he left for publication I have been greatly assisted by Professor R.F. Tylecote and Dr Lynn Willies.

The author's draft includes considerably more material than has been published here. Some of the additional material concerns the RTC copper smelting operations in South Wales, there is more information on the ancient slags, a significant section concerning the developments in sulphuric acid production, and more social history.

In the expectation that publication of this edited version could stimulate interest in the further information, a copy of the draft is lodged in the Institution of Mining and Metallurgy and can be examined by arrangement with the Head of Library and Information Services.

* Rua Figueroa Ramón. *Minas de Rio Tinto: estudios sobre la explotación y el beneficio de sus minerales* (Coruna, 1868).

For more detailed archaeological information the reader is referred to the publications of the Institute for Archaeo-Metallurgical Studies attached to the University of London, 31-34 Gordon Square, London WC1H 0PY, England, and to its volume *Studies in ancient mining and metallurgy in South-West Spain.**

Most of the old plans and photographs used to illustrate this volume have been copied from archives in the possession of RTZ Limited and from publications of the Institution of Mining and Metallurgy; exceptions are acknowledged in the text.

The Institution of Mining and Metallurgy is grateful to RTZ Ltd for providing the financial support needed to enable it to undertake publication.

M J Cahalan
44 Portland Place
London

January 1987

* Rothenberg B. and Blanco-Freijeiro A. *et al. Studies in ancient mining and metallurgy in South-West Spain: explorations and excavations in the Province of Huelva* (London: IAMS, 1981), 320 p.

Introduction

In the south-west of the Iberian Peninsula there is a vast pyritic mineralised zone, known as the Andevallo, extending from near Seville to south of Lisbon, an area some 150 kilometres long and 30 kilometres wide.

The Rio Tinto Mines, which are the largest of this "pyrites belt", lie in the region known as Andalusia, some 90 kilometres north-west of Seville and 75 kilometres north-east of Huelva. They have a very long history, dating back to pre-Iberian times; then came the Iberians, a race of North African origin (Turdetarian and Tartessian), the Phoenicians, the Carthaginians, the Romans, the Moors, the Spaniards and the British. It is believed that copper was first recovered from the ores in the third millenium BC and that silver was mined in the late Bronze Age, 12th to 9th centuries BC onwards.

The Phoenicians established a trading station at Gades (Cadiz) about 1000 BC and found the hinterland rich in both agriculture and metals - gold, silver, copper, tin, lead and iron. Their efforts to dominate the country were frustrated by the resistance of the Tartessians living to the north-west but they apparently traded successfully and are believed to have used the fabled port of Tartessos. The exact location of Tartessos has not been identified but it was probably on the estuary of the Tinto and Odiel rivers near Huelva.

In 535 BC, with the decline in power of the Phoenicians, the Carthaginians conquered southern Spain. They too traded in agricultural produce and metals and also experienced difficulties with the Tartessians. They were responsible for the destruction of the port of Tartessos.

Whether the Phoenicians and the Carthaginians ever actually worked the mines, or were merely traders, is not certain, but after 205 BC, when they defeated the Carthaginians, the Romans brought their own men skilled in mining and metallurgy.

The Romans occupied most of the Iberian Peninsula for 600 years, until about 425 AD - the most recent Roman coins found at Rio Tinto show the head of Honorius who was emperor from 395 to 423 AD. Mining must have declined with the invasion of Barbarians in the 5th century and the subsequent entry of the Visigoths who were eventually absorbed into the people of Spain. In 711 AD the Moors invaded the Peninsula from North Africa. Little mining was done after the departure of the Romans; the Moors probably recovered some copper from acid drainage waters by deposition on iron.

The Moors were driven out of Spain in 1492 and the first extant report on the Rio Tinto mines is dated 1556. In 1725 a Swede, Liebert Wolters Sjohjelm, was granted a licence to work five mines in the region, one of which was Rio Tinto - the documented history of operation of the Rio Tinto Mines begins from that date.

Over two and a half centuries the mines were worked both directly by Government and by private companies. In general, Government operations were unsuccessful, despite the outstanding men who served Government as adminstrators at Rio Tinto. The Treasury in Madrid was reluctant to supply funds for desirable investment, for example when steam power became available it was not installed; miners could not work lower than the Roman adits, yet steam powered pumps could have resolved dewatering problems.

In April 1867 a Royal Commission made recommendations to develop the mines but the Treasury was, as ever, reluctant to fund the recommended programme and the Government decided to sell the mines. During a period of great political turmoil, an authorising law was passed by the Cortes in June 1870 and bids were invited in the following year. An offer by a consortium of British and German bankers was accepted and The Rio Tinto Company Limited was registered in London in March 1873. The events leading to the purchase by this new company have been chronicled in detail in David Avery's "Not on Queen Victoria's Birthday"(1).

The purchase price for the mines, in the form of the freehold of an area of some 1,900 hectares, was £3,600,000 equal to 92,800,000 pesetas. This was a huge sum at that time and, although slightly below the Spanish Government's valuation, was regarded by others who had contemplated purchase as excessive.

The Chairman of the Rio Tinto Company, Mr Hugh Matheson, clearly had great faith in the reports he received on the mines from Dr Roemar, a German mineralogist, Mr David Forbes, a British mining engineer, and Messrs Sundheim and Doetsch, general merchants operating from Huelva with interests in mining. Doetsch was the catalyst who brought German finance to join Matheson's marketing and management experience.

It certainly did not appear to be a propitious time to make such a massive investment in Spain - the State was bankrupt, a Republic had just been declared and the country was in a state of virtual anarchy. However there was a rapidly growing demand for pyrites for sulphuric acid production in Europe for the expanding chemical industry.

The Rio Tinto Company continued mining and smelting in Spain, through two world wars and a civil war, economic booms but more long lasting depressions, until 20 June 1954 when two-thirds of its Spanish assets were sold to a Spanish group of bankers and the company was renamed "Compania Espanola de las Minas de Rio Tinto".

Few mines have such a history as Rio Tinto. These notes concentrate on technical aspects of mining and metallurgical activities using information obtained from a wide variety of sources, some well documented, some anecdotal. There is inevitably a degree of speculation about many aspects but, on the whole, we can deduce a fairly convincing picture of the activities which have taken place over more than 4,000 years.

It will simplify later references to locations to give at this stage some general description of the mining area as it has taken shape in recent, recorded history. Figures 1 and 2, taken from the prospectus for the sale of Rio Tinto Company shares to the public, give details of the local geography.

The Rio Tinto Mines (Minas de Rio Tinto) are situated in the western foothills of the Sierra Morena, in an area sometimes referred to as the Sierra de Aracena, with the main lodes divided by a ridge which contained four peaks - Colorado, Salomon, Retamar and San Dionisio. Between Retamar and San Dionisio is a pass called Puerto Rubio. The highest peak, Cerro Colorado, was 534 metres above sea level but has now been lost through opencast mining. The most impressive peak was Cerro Salomon, 515 metres, but this too will soon be mined away. Earlier writers about the area thought that the name Salomon came from the biblical King Solomon who came to the throne of David about 970 BC; however the earliest recorded reference to Cerro Salomon comes from the 17th century.

The British company began its operations in 1873 with opencast mining of the South Lode, followed by underground mining at South Lode and

San Dionisio in 1880. In 1883 a start was made on opencast working of North Lode and three further opencast mines came into operation at the beginning of this century - Lago in 1903, Dehesa in 1906 and Atalaya in 1909. Since Compania Espanola de las Minas de Rio Tinto took over in 1954 major opencast mining of gossans for gold and porphyry for copper has transformed the landscape.

Mining from 1873 relied on rail transport to and from the port of Huelva, and for mineral, waste and supplies movement in the mining area. Installation of the railway was a major commitment for the new company and, seen in retrospect, railway development added to the opportunities to uncover early mining and metallurgical activity.

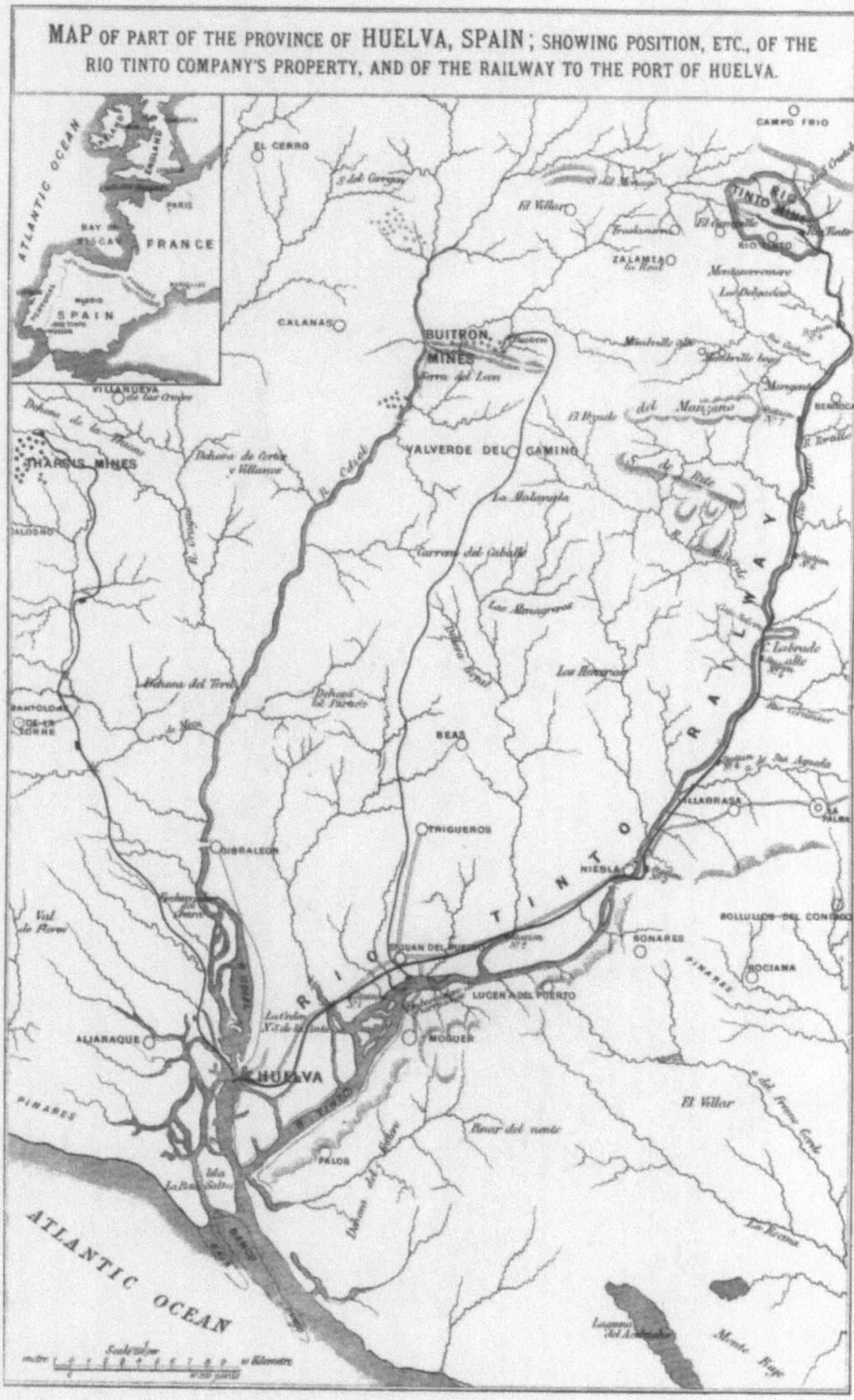

Figure 1

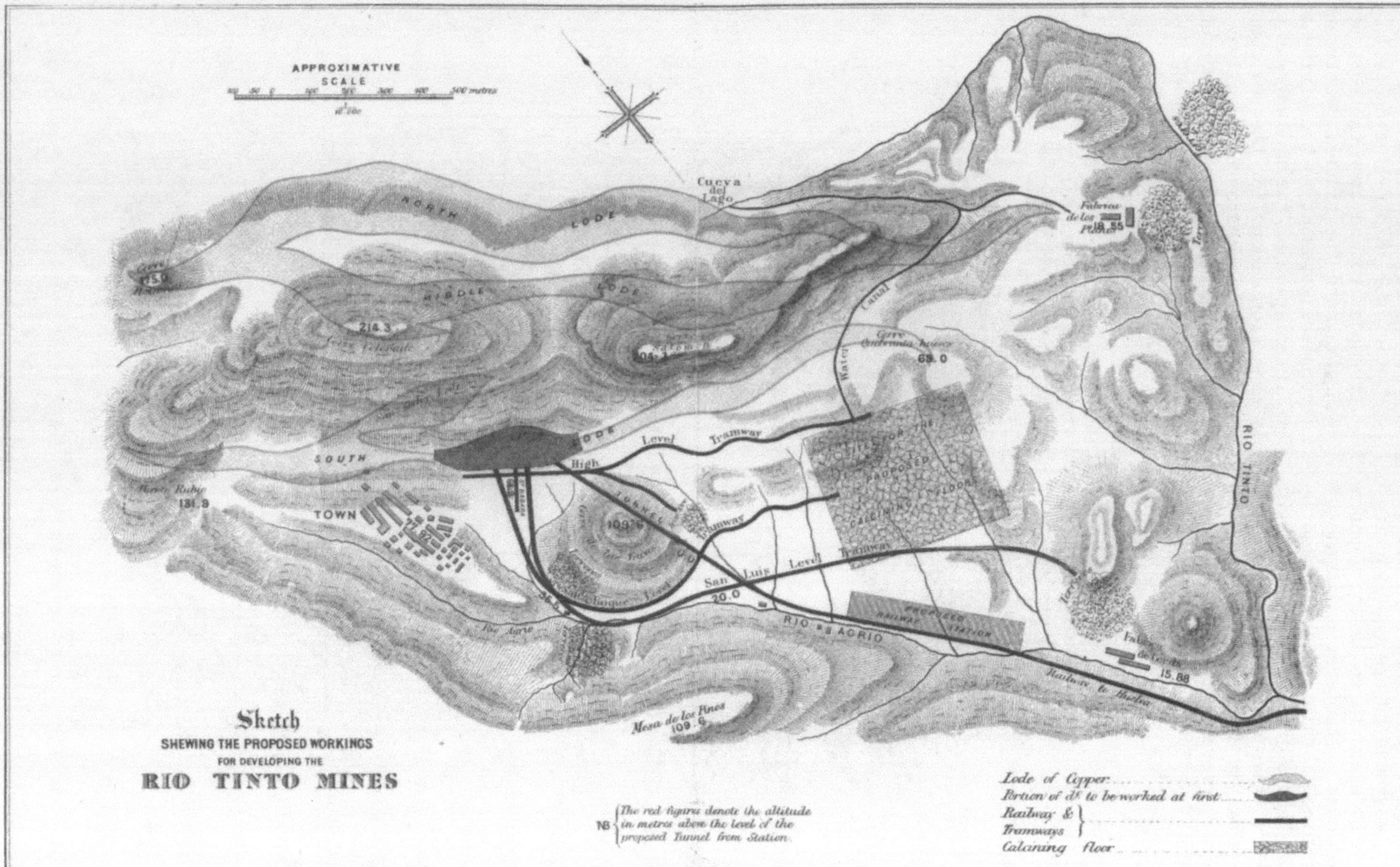
APPROXIMATIVE
SCALE
500 metres
Sketch
SHEWING THE PROPOSED WORKINGS
FOR DEVELOPING THE
RIO TINTO MINES
NORTH
LODE
MIDDLE
LODE
SOUTH
LODE
Cueva del Lago
RIO TINTO
RIO AGRIO
TOWN
High Level Tramway
San Luis Level Tramway
TUNNEL
PROPOSED CALCINING FLOORS
PROPOSED RAILWAY STATION
Railway to Huelva
Water Canal
218.3
204.1
150
131.9
109.6
69.0
20.0
18.55
15.88
Mesa de los Pinos 109.6
NB The red figures denote the altitude in metres above the level of the proposed Tunnel from Station.
Lode of Copper
Portion of d° to be worked at first
Railway & Tramways
Calcining floor

Figure 2

CHAPTER 1 The pre-Roman period

Native copper was probably the first discovery at Rio Tinto by man using metals. It would be produced from the copper sulphate in acid drainage waters by reduction with sugars from decaying vegetable matter. Such native copper would have been found in filamentous or lace-like form, requiring consolidation into artefacts, such as pins and fish hooks, by hammering and possibly heating.

The earliest metallurgical activity in Spain was probably that of the Western Neolithic culture in the province of Almeria, some 350 kilometres to the east of Rio Tinto. This area was settled by people with crafts and habits similar to those of the people of the Nile valley. By 2,500 BC they were extracting copper.

Ores used in this copper production were probably basic copper carbonates and the technique for smelting such ores may have spread fairly readily to the west. An alternative possibility advocated by some archaeologists is that production of copper in Andalusia may have been indigenous and developed earlier than in Almeria. There is little doubt that basic copper carbonates, produced by secondary enrichment from the cupreous pyrites, would have been available.

Methods used for the reduction of copper carbonate ores in antiquity have been widely studied and there is now a clear understanding of the techniques used. Furnaces, the remains of which consisted of a hole in the ground lined with clay, were charged with ore and charcoal. Combustion air was supplied initially by blowpipe but the use of bellows from about 1,200 BC was an important advance. The smelted charge was cooled in the bowl, after which metallic copper was obtained either as prills by crushing the mass of slag, or as a "bun" where the smelting temperature and slag fluidity were sufficiently high.

Analysis of copper carbonate ore found in recent times at Rio Tinto gave:

Cu	11.30 %	As	0.054 %
Pb	0.28 "	Sb	0.054 "
Zn	0.01 "	Bi	0.043 "
Sn	0.006"	S	trace
Au	0.20 grams per tonne		
Ag	285.0 " " "		

It is probable that the ancient workers, by careful selection of what would have been mineral specimens, were able to feed considerably higher grade mineral to their smelting furnaces.

Recent archeological work has confirmed that Rio Tinto was a major source of silver in ancient times and it is interesting to speculate on how the original discovery of the silver ores was made, especially as for the first twenty years of mining by the Rio Tinto Company the silver rich jarosites were discarded in opencast mining.

Copper ores were, of course, readily identifiable by their bright green and blue colours but the silver ore, argento-jarosite, is usually described as a "coloured earth" and the discovery of its value as a source of silver could probably only have been made by heating it, alone or mixed with charcoal. The source of the Tinto river was apparently in a cave in an area known as Lago where the coloured earths occurred and it is assumed that it was here that the ancients made their discovery.

Neil Kennedy, a mining engineer employed by the Rio Tinto Company from the beginning of its operations, wrote in 1893(2):

> "When the open-cutting of the South lode was being made a stone was found 14 metres below the surface (in a cave) in the exact line of contact between the iron pyrites and the capping which superimposes it. This capping is generally a coarse iron ore, but in this particular spot where the stone was found, it was formed of an ochrish looking earth. The porphyry stone was 38 cm long by 30 wide and 11 cm deep with eight holes cut in its surface, each 7.5 cm diameter and 3 cm deep, accompanying it was a pestle also of porphyry which had a rounded head to fit the circular holes. Another very similar stone was found some little distance off."

A geologist who examined the stones thought the primitive inhabitants must have employed them for mixing pigments for body decoration. It must have been soon after this discovery that it was again realised that the "ochrish-looking earths" were silver ore.

Lead is an essential requirement in the recovery of silver from ores by smelting but it is generally assumed that there was little at Rio Tinto and elsewhere in the region, and the jarosites found at Rio Tinto are usually low in lead. J C Allan(3), writing in 1970, referred to this deficit of lead in relation to the role of the Phoenicians:

> "Even if there is no archaeological evidence of eastern Phoenicians' influence earlier than BC 800 this does not seem to be a valid reason for supposing there was no export of silver earlier than this date...
>
> "Interesting in this connection is the comment by Timacus c. 400 BC that the Phoenicians on their return journey substituted silver anchors for the usual lead ones ... Thus this .... can be considered as the record of a perfectly normal commercial transaction in which the lead anchors were traded with the silver producers in return for their product."

If that interpretation is correct, the Phoenicians had a remarkably well developed approach to optimising transport costs and to the provision of lead for the silver producing operations. In cupelling, the classical process for obtaining silver from the lead bullion in which it is concentrated in smelting, a considerable amount of lead is lost, quite apart from the oxidation of the lead to form litharge and that absorbed in the cupels. Incidentally the smelting to bullion was probably accomplished in primitive furnaces with charcoal fuel as indicated above for carbonate copper ore.

Although many had believed that silver, gold and electrum (the alloy of the two) had been produced at Rio Tinto in very ancient times, the first convincing evidence was established in 1966 by archaeologists from the University of Seville(4,5). On the south side of Cerro Salomon they found foundations of a village extending along the side of the ridge for a total length of 900 metres. Water for this village, high above the surrounding terrain, was obtained from a spring in the saddle between the two summits, Cerro Colorado and Cerro Salomon.

The houses consisted of collections of small rooms with walls made of undressed stone; the period of occupation has been dated as c. 800 BC. Stone pestles and mortars were found and the earth floor contained stringers of lead which had leaked from a smelting hearth. Numerous clay tuyeres were found and the slag on some of them had a green colouration, as though they had been used in smelting copper ore. No furnaces were found, but in one room there was a hearth, about a metre in diameter and 40 cms deep, containing partly calcined bones and ashes - could this hearth, and others like it found in the Lago area, have been used to smelt charges in crucibles? And did these ancients know that bone ash was the best material for cupels?

A considerable quantity of shards of the Phoenician period was found and a large amphora standing on a piece of grey slag. This slag was unusual - it was of greyish colour and obviously only semi-fused, with many holes and embedded silica particles of the type known locally as "guijarro". The silica particles ranged in size between 5 and 15 mm and the silver content of the slag was high - 575 gms per tonne. Earlier reports, of Courtney(6) and Gonzalo y Tarin(7), mention slags of this type. Gonzalo y Tarin mentions seeing conical mounds of wrinkled, sponge-like slag, greyish in colour and imperfectly melted, in association with well fused black slag, at some mines in the province of Huelva.

More significantly, Courtney wrote:

> "There is no distinct evidence of the mines having been worked about this time (500 BC), although the fact of several heaps of a rough and spongy class of scorial being in close proximity to the lodes, and always near abundant spring water, has strengthened the belief that the heaps are the refuse of Phoenician smelting."

As so often happens, after the importance of this slag type had been recognised following the 1966 discovery at Cerro Salomon, other samples have been identified. One was found in the museum at Niebla, a village on the site of an old Roman town on railway between the Rio Tinto mines and the port of Huelva. This specimen was described as follows:

> "Slag found loose in the soil in the course of the excavation of the "Campi", but never in the cave retorts at Niebla. Similar samples of slag given to the museum some years ago, with the information that it had been found on the by-pass leading up to the Sierra of Aracena, but its exact position could not be explained. Two or three samples were found like those in the cyclopean wall on the Desembarcadero (wharf) in 1924."

The Niebla museum was founded by Mrs Elena Whishaw, the widow of an American archaeologist who lived in Seville. After her husband's death she moved to Niebla and built a house in the old Moorish walls. She was a friend of Lord Milner, who was Chairman of The Rio Tinto Company from 1923-25, and was much respected by the local people who she employed in continuing archaeological excavations and in producing sand for sale. She was responsible for the discovery of the ancient port of Niebla.

It seems probable, based on the Niebla slag specimen, that smelting of silver ore was also carried out there. Several specimens of the poorly fused slags from around Huelva, Niebla and Rio Tinto mines have been analysed in recent years and show low copper and lead contents, with total silica contents over 40% and silver contents of several hundred grammes per tonne.

Figure 3 Niebla - Walled Roman Town

In 1973, a slag of this type found at Calle Palos near Huelva was dated at 400 BC by thermoluminescent dating. A portion of the same specimen was examined to establish its fusibility - at 1000°C the edges of the specimen were tending to round; at 1200°C the specimen was becoming plastic and changing shape; it was only when the temperature reached 1400 C that it became molten. Such a temperature could not have been obtained in a primitive furnace and no satisfactory explanation is yet available for apparently adding silica particles in such large quantities to the fused mass.

Elucidating pre-Roman activity on the basis of examination of slags has made major progress in recent years. The vast tonnages of slags from Roman exploitation created considerable difficulties which were further compounded by the superimposition of workings after the rediscovery of the mines in the 16th century.

Exploitation by the Phoenicians created some infrastructure and references from early literature can be pieced together to suggest the identity of some of the places named in Greek and Biblical sources.

Silver metal, perhaps in the form of bars weighing between 4 and 15 minas (a mina was a weight of about 400gms) must have been collected from mines, large and small, and sent to depots from which it could be delivered to a port capable of handling sea-going vessels and providing shelter from severe weather. There may also have been movement of jarositic ore, from small mines to larger mines with smelting facilities, or possibly for export, although that seems less likely.

In the province of Huelva there were two places suitable for depots; one was Niebla which, on the evidence of the excavated port, would have been capable of taking the large boats. The other possible site is Araraque, on the west side of the estuary of the rivers Odiel and Tinto and opposite the present town of Huelva.

The historian Herodotus (c. 484-425 BC) wrote about Colaeus of Samos whose "bajel" was blown through the Straits of Gibraltar in a severe gale to landfall at the port of Tartessos. This has generally been assumed to have been near the mouth of the river Guadalquivir but a much more logical location would be in the Huelva estuary. Luzon(8) supports the Huelva hypothesis and identifies the site with the biblical Tarshish - at least two passages in the Bible(9) state that ships came to Tyre from Tarshish every three years bringing, among other items, gold and silver. Bronze items, dated at c.700 BC were dredged from the Huelva estuary in 1923 and are on display in the museum in the town.

The Greeks wrote that Tartessos was on, or near, a river whose source was a cave in a silver mountain. This could have been the Tinto river and the Lago cave. However, early reports of this river mentioned that particles of tin were seen in it; there is no tin-stone or cassiterite in the Tinto river but there could have been grains of ilmenite, whose colour and appearance are not very different from those of cassiterite. Before the Romans opened up the mines and thus provided the conditions to promote oxidation of the pyrites which was the first stage in producing the red coloration of the drainage waters, the river would have been only slightly coloured.

By 400 BC the port of Tartessos was already in decline, perhaps because of silting or because the island on which it was built was sinking. It was finally destroyed by the Carthaginians.

There was probably another depot further west on the Guadiana river, near the present day town of Ayamonte, where tin or tin ore was stored. Alluvial cassiterite was certainly to be seen in the bed of the Guadiana and has been observed there in recent times.

CHAPTER 2 The Roman period

When the Rio Tinto Company Limited (hereafter normally abbreviated as RTC) started its operations in 1873, there must have been some evidence of mining prior to entry of the Romans. Unfortunately, little notice seems to have been taken of such evidence, the staff being fully involved in the exciting new activities, and it was probably obliterated. In any case interest in pre-history was not at all common in the 1870s.

The evidence of Roman occupation was, and still is, much more substantial, notably in the extensive slag heaps, although there has been some exaggeration about the size of such heaps. Diego Delgado, in the first extant report(10) on the Rio Tinto mines dated 1556, wrote:

> "These slag heaps are so massive that they appear to be large hills and mountains. Could these slag heaps which we saw up to two leagues long and two leagues wide without doubt be those which we were told were more than eight leagues long?"

Again, in 1632 Caro(11) wrote:

> "Near the mine can be seen hills of charcoal and slag for many leagues, a simple glance gives a terrorizing aspect causing apprehension and dismay to the casual and infrequent visitor. For leagues it covers the ground and rises to hills."

Since a Spanish league of that time was about 5.5 kilometres, in the light of what we have now established of the Roman operations, both the above authors must have taken considerable licence in their descriptions. Later, at the time of the offer of the mines for sale in 1871, the official inventory stated: "The amount of slag accumulated in the place called Dehesa is incalculable" and no doubt the calculation would have proved a daunting one with the facilities available to those compiling that inventory.

Not much is visible of those slag heaps today; they have largely been covered with overburden from recent, post 1960, open-cast mining. From the evidence of the coins found, many in the slag heaps, it has always been assumed that the Romans occupied the Mines from the time of Augustus (27 BC to 14 AD) to that of Honorius (395 to 423). A sample of charcoal taken from a slag heap at 2 metres above ground level and under a further 4 metres of slag gave a carbon 14 dating of 140 AD + 95 years which fits reasonably the assumed time frame but some authorities now believe the period of occupation was somewhat longer, perhaps as much as 600 years.

Strabo(12), who lived from 63 BC until 19 AD, indicated the position of the mining field. He wrote:

> "A chain of mountains parallel to the Boetis (Guadalquivir) extending towards the north .... they enclose a quantity of mines .... Near the site called Cotinas is produced gold and copper. These mountains are to the left of those which ascend the river. .... The Anas (Guadiana) is also navigable, but not for such a length, nor with such large vessels. Its edges are also bordered by mountains which contain mines."

Some people believe that Cotinas is Rio Tinto; certainly the mountains enclosing Rio Tinto can be seen from Italica near Seville. The probable importance of the Guadiana as a mineral route for tin, copper, gold and silver has been referred to in Chapter 1.

In Roman times the Tinto river was named Urium, meaning burning, because, presumably, of the corrosive nature of its water. Its source was said to be in a place named Gran Bitania and the port at its mouth, now Huelva, was known as Onuba.

There are several accounts of the mine workings provided by visitors after the rediscovery by Diego Delgado. According to Gonzalo y Tarin(7), who apparently produced a geological survey of the area in the 1880s, soon after RTC began exposing old workings in the course of its operations, there were no less than thirteen Roman adits; nine to the east of Cerro Salomon and four to the north.

The Romans worked the jarosite formation for silver and the main sulphide bodies for copper. The exposed area of green malachite and blue azurite might, according to some geologists, have always been quite small. However, Gonzalo y Tarin(7), wrote:

> "The ancient mining consisted of shallow opencast workings following the direction of the ore. The very similar disposition and layout of all these workings presupposes that the ore outcropped, which naturally helped the miners to find the best places for extracting the ore."

He was reporting on the many small opencasts that he saw at a number of mines in the province of Huelva where, in the main, copper carbonate ores had been extracted

In 1856, Anciola and de Cassio(13) reported seeing opencasts on the spur north of Cerro Colorado, at Dehesa, south of San Dionisio and at other places and concluded they had been made by Romans. Rua Figueroa, who was resident at the mines from the 1850s and manager from 1859 until 1863, also refers to such opencasts in his publications(14,15). Of particular interest is a report of Lee Thomas(16). He visited the province of Huelva in 1865 and has provided one of the few detailed impressions of Rio Tinto prior to the commencement of operations by RTC. Two extracts from his report are relevant here:

> "In many mines of the district, amongst them El Tharsis, Buitron and Rio Tinto, I observed a very marked depression in the ground immediately over the deposit, the walls of which rose above its level and defined clearly the width of the mineral below. Where these depressions are observable, I myself have no doubts that the Romans, or their predecessors, worked the mines by means of open-cutting. This view is confirmed by the character of the filling in the cases in question, the oxide of iron (gossan) is found in detached masses, and the decomposed porphyry and slate predominate. The line of junction between the covering and mineral is regular and unbroken and looks more like the work of human hands than a natural formation. At the Lagunazo mine, where works have lately been commenced by an English company, ore has been cut to a depth of seven metres from the surface; this I am inclined to believe is no special instance, and that in cases where the depressions described above are observed, the ore has been found near the surface and taken away by old men. In many instances old shafts are found perforating the filling-in which covers the mineral. These I am inclined to regard as the work of Romans, and that when they came to Spain they found the backs of the lodes had already been taken away by the Phoenicians and Carthaginians and were compelled to have recourse to shafts and adits to continue the working of them."

> "In the vicinity of most of the deposits of pyrites, there are quantities of old slag, but at Rio Tinto the extent of the ground covered with them is immense; there are hills composed entirely of

> them, which literally cover a considerable surface of country; the quantity appears fabulous, and must be seen to be credited .... In comparison with Rio Tinto the mounds of slag to be seen at the other mines of the district are insignificant, and almost lead one to the conclusion that the former must have been a centre of smelting operations, on account of the abundance of fuel in the vicinity. This theory is rather borne out by the occurrence of lead slags at Rio Tinto."

Despite the impression made on him by the vast accumulations of slag, Thomas was "struck with the seeming insignificance of the Roman workings, in those parts of these mines recently opened, compared with what I had expected." This would have reinforced his supposition that Rio Tinto had been a smelting centre; as became clear in later years and was probably not obvious at the time of his visit, underground operations had been conducted in several areas and the full extent of them was not apparent until the RTC operations were well advanced.

Kennedy(2), wrote shortly after RTC began operations:

> "The opening of these mines in modern times has shown that the ancients simply worked the richest spots .... Now these rich pockets were not easily or quickly found by a race who had no knowledge of blasting operations and although the main masses of pyrites are, as a rule, easily discovered, yet those of themselves were not rich enough for the ancients"

There would have been some secondary enrichment and Rua Figueroa(15), reported finding chalcocite containing 68 % Cu near a mine shaft. That was no doubt very exceptional; it is usually assumed that the Romans smelted ore of between 10 and 22% Cu.

John F Allan, another of the early mining engineers with RTC, reported in 1887(17):

> "The Roman mining works are on a gigantic scale, considering the means at their disposal, adits at various levels are still open and large cavities are found. The largest workings are set near the walls of the lodes where the copper minerals are richest and easiest to work. The upper levels of the mines contain secondary enrichment and chalcocite was found near the walls of the lode."

Adits were employed to drain the operations, a relatively simple matter for the jarosite mining operations but copper mining extended below adit level and water wheels were used to lift water. Some thirty such wheels have been found in the workings at Rio Tinto, the largest 4.6 metres in diameter. The hubs of the wheels were made of oak, the spokes, buckets and rims of pine, and the assemblies fitted together with wooden pegs and mounted on a bronze axles. Motive power was provided by men standing on cleats attached to both sides of the wheels.

In 1919-20 in the Planes mine, a rich copper area of Rio Tinto which had been extensively mined by the Romans, a set of eight such wheels was found. This installation was arranged to lift water almost 30 metres. Palmer(18) has written an account of some of the old workings and of this discovery - he tested a reconstructed wheel and established that it needed a force of about 70 kilos to operate, raising water about 3.7 metres.

On the Roman mining methods Palmer wrote:

> "Although the orebodies were penetrated by many ramifications of small galleries, the author knows of no large stopes having been found, and thus he can produce no detailed evidence of the general mining methods adopted. In the oxidised zone, from which apparently the ancients obtained most of

the minerals mined, and where the overlying burden was more or less shallow, the system adopted appears to have been that of sinking a shaft, or possibly a pair of shafts, into the zone, extracting all the ore which could easily be reached from these shafts, and then sinking other shafts in the immediate vicinity.

"When the overburden, however, was greater than the shaft system warranted, the system appears to have consisted of providing timbered galleries which radiated in all directions, and extracting what could be reached from these. This system was apparently very effective, as not much of any value appears to have been left in those places where the gallery system was complete."

Figure 4 Reconstructed Roman Water-lifting Wheel

The significance of the operations for silver was not appreciated by Gonzalo y Tarin. It is true that the opencast operations were the source of a large part of the copper ore smelted by the Romans and the Phoenicians; we now know that a large fraction of the slags were the result of silver production from the jarosite formation and most of the shafts which early 19th century visitors saw were used in mining the jarosite silver ores.

It is easy to understand why those examining the mining areas of the province in the 17 - 19th centuries did not appreciate the significance of the exploitation of the Rio Tinto deposits for silver; the obvious association was with treatment of copper. There was continuing mystery associated with the very low copper content of much of the slag - that derived from silver recovery - and other evidence of silver recovery was sometimes assumed, for example by Rua Figueroa(15) to indicate only that the Romans recovered the silver from the copper; they were known to have done that elsewhere.

However, Fernando Bernaldez, who was employed at Rio Tinto as a metallurgist, wrote in 1853(18) putting forward the view that the slags at Dehesa arose from smelting lead ore. He found metallic lead containing about 1700 gms per tonne of silver in the Dehesa heaps, whereas copper found on what he considered to be slag heaps from copper smelting, contained only the same amount of silver as that in the copper ore, say about 50 gms per tonne. Because of these investigations he wrote: "So we are not making foolish statements when we say that the original miners recovered silver and gold."

In recent times, the location of some slags, clearly from copper smelting, underlying slags from silver production was a puzzle until 1979, when a team of archaeologists under Dr P T Craddock, found a very large accumulation of copper slag, much of it in the form of large discs such as would have been formed in tapping the copper and slag into a bowl. The significance of these slag discs has been described by Tylecote(20).

The location of this slag, which was far removed from the main sulphide bodies, is now assumed to indicate that it was derived from the smelting of carbonate ores. Although geologists earlier this century were of the opinion that the original formations contained little such oxidised ore, this find supports the views put forward by Gonzalo y Tarin and Lee Thomas that significant exploitation based on opencasts had taken place in ancient times - by Tartessians, and possibly Phoenicians and Carthaginians, as well as Romans. It now seems that the Romans did continue to work oxide deposits where they were available, but their major technological achievement was in underground mining of the sulphide copper and jarositic silver ores.

It is important to understand a little of the geology to appreciate the significance of the silver operations. Professor David Williams spent many years at Rio Tinto from the late 1920s and published several papers on its geology. From one of these publications(21), the following extract has been taken:

> "The thick mantles of gossan formerly overlay the massive deposits (of pyrites) at San Dionisio, South Lode, North Lode and Planes, whilst the conspicuous reddish gossan crown of Cerro Colorado represents the oxidised remains of long vanished ore bodies. The base of the gossan is generally flat lying and is usually more sharply defined when resting upon massive pyrites than when covering heavily iron stained porphyry. Over a wide area of Cerro Colorado the gossan, for a thickness of about 24 metres, contains the equivalent of more than 50% iron together with 1.25 oz (43 gms per tonne) of silver and 1 dwt (1.7 gms) of gold. Both the arsenic and lead contents of the massive gossan are approximately 0.7%; copper averages 0.1%. Jarosites do not appear until the base of the gossan is reached ..... the lower part of the gossan is commonly marked by an earthy layer, up to 1.25 metres thickness, which is compararively rich in gold and silver, lead, antimony (arsenic), bismuth and selenium. Locally, this layer includes bands of yellow, red, grey and black earths whose relative positions are not always constant.
>
> "In most places, the yellow variety predominates, often to the exclusion of others .... in hand specimens the yellowish material might easily be mistaken for ochreous limonite, or some iron stained earthy material. Much of it can be crushed with the fingers."

David Williams estimated that 3 million tonnes of jarosite had been present before mining began and that perhaps 2 million had been extracted by the Ancients and the Romans. However, the vast tonnage of slag found at Rio Tinto, until recently estimated at 20 million tonnes, indicated the smelting of a considerably greater quantity of ore. Calculations indicate that about 4.5 tonnes of slag would normally be produced in smelting one tonne of ore so that 2 million tonnes of jarosite would yield 9 million tonnes of slag. To this should be added an allowance for resmelting of high silica Phoenician slags and for smelting ores brought to Rio Tinto from other mining sites, still suggesting a large discrepancy.

Although RTC was not immediately concerned with unravelling the early history of operations at the site, a survey of the old slag heaps was made in 1880 and a further survey in 1924; meanwhile large quantities of the slags had been used as ballast on the extensive railway system. It is estimated that as much as 4 million tonnes was used for railway and other construction purposes before the 1924 survey which listed 1 million tonnes of copper smelting slags assaying 0.74% Cu and 4.7 gms per tonne Ag, and 15 million tonnes of silver slags assaying 0.13% Cu, o.40 % Pb and 54.3 gms per tonne Ag.

The figures for copper slags include slags made in the 18th and early 19th centuries. A significant tonnage of copper slags was buried by 18th century operations and hence not included by the survey. [Since Leonard Salkield's death, a new survey of the ancient slags has been completed under the direction of the Institute for Archeo-Metallurgical Studies and this has established a much lower figure of around 6 million tonnes which accords quite well with the smelting of some 2 million tonnes of jarosite plus lesser amounts of copper ore - Ed.]

The calculation of a slag make of 4.5 tonnes per tonne of silver ore smelted is based on an assumed 50-60% silica in the jarosite ore. In fact the jarosites found at Rio Tinto and analysed since 1893 show wide variations for all key constituents, as follows:

Silica	0 to 92%	Bi	tr to 0.25%
Fe	2 " 33%	Barytes	0 to 9%
Cu	0 " 0.16%	CaO	0.2 to 0.5%
Pb	0 " 40%	MgO	tr to 0.5%
Sn	0.05 " 1.7%	Mn	" " 0.12%
As	tr " 36%	Alumina	0.7to 2.5%
Sb	tr " 5%		
Au	2 to 50 gms per tonne		
Ag	160 " 6800 " " "		

In 1933, the RTC Mining Department Annual Report referred to an investigation of old Roman workings in exploring the gold-silver layer at the base of the gossan.
F N Spettigue wrote:

> "The area between South Lode and Dehesa is full of old Roman workings and it was decided to explore these workings to find further deposits. A Roman shaft in this vicinity was cleared out and the extensive workings at the base of the gossan examined.
> Some of the ironstone filling ran 8.4 gms Au and 198.8 gms Ag per 1000 kilos. It looks as if the majority of the ore, in the immediate vicinity of the shaft, has been mined out by the Ancients ....".

Between about 700 BC and the entry of the Romans to Rio Tinto, the primitive batch furnace slowly developed into the blast furnace. With this advance, it was no longer necessary to allow the furnace to cool before removing the slag and metal, instead the molten materials could be discharged into pits, allowing the furnace to receive another charge while still hot.

Rua Figueroa observed that the Romans usually built their furnaces on top of mounds, as near the source of ore as possible, and, because of lack of water, the bellows were hand operated. This type of furnace was used to smelt both copper and silver ores and it remained in use in Spain until the middle of the ninteenth century. John F Allan(17) also noted details of Roman furnaces he saw at Rio Tinto in the 1880s:

> "One side appears to be formed by excavating in the solid rock, with a semi-circular wall about 7 feet high built in front of it, leaving a circular section 2.5 feet in diameter, with openings below for the overflow of the slag and the admission of blast."

The fuel used was charcoal, produced from the indigenous evergreen oak, quercus ilex. Smelting basic copper carbonates with some chalcocite was relatively straightforward; there would not have been any need to remove sulphur as there was sufficient copper oxide to react with the sulphide during smelting.
Neither would smelting of silver have presented serious problems as most of the lead was present as sulphate, easily reduced by charcoal to produce the lead-silver alloy.

Smelting galena was more difficult, probably involving roasting to oxide because the sulphide is not easily reduced by charcoal. Possibly it was insufficient roasting of galena which, in Roman slags, is associated with high lead contents. Curiously, the slags at Rio Tinto are low in lead compared with slags at Laurion, up to 10% Pb, Cartagena, 8 to 17%, and Sardinia, up to 30%. One of the reasons why there was some reluctance to accept that so much of the slag at Rio Tinto arose from lead-silver smelting is its low lead content, average around 0.4%.

The deficiency of lead at Rio Tinto, and in the nearby areas, has been referred to in Chapter 1. The Romans used lead extensively for plumbing and had well established producing centres. It is recorded by Rickard(22) that two ingots of lead, marked NOVA CARTHAGO, were found in a slag pile and the inference is that the Romans brought lead from Cartagena for the metallurgical operations. [Rickard, in a written contribution to the discussion of Palmer's paper (18), referred to export by RTC of "several thousands of tons of ore containing 34% Pb, 1.22% Sn" with no indication of whence it came. Salkield verified this as an 1895 shipment of silver ore - plumbo-jarosite - containing 1460 grams Ag per tonne. In all RTC operations, the total quantity of such jarosites extracted is thought not to exceed 10,000 tonnes. Ed.]

A considerable collection of Roman tools has been assembled from the workings. Iron tools have been found in the jarosite workings, in slag heaps and in the houses of workers, but not in the sulphide workings where oxidation led to acid copper solutions; these dissolve iron in the cementation reaction. In common with other contemporaneous non-ferrous smelting operations, the Romans are presumed to have had a smelter to produce iron from gossan somewhere nearby but the site has not been identified.

The establishment by the Romans of substantial operations and a large resident community required assured water supply in a relatively arid area. This was obtained from springs and also by sinking wells in the slate, applying the recognised expertise of the Romans in water engineering. These same sources of supply were later used well into the 18th century. One of the Roman installations to the northeast of Cueva del Lago was described around 1900 by William Nash(23):

> "It consisted of an adit driven into the hill some 140 yards, a cross-cut was put in at every 15 yards, each arm extended 8 yards, at each of these junctions a well was sunk and so on regularly through the whole length of the adit. The work is quadrangular in form, and has the reputation of being Arabic in origin; but probably the Arabs only bettered the condition of some much older work. Of the few permanent and substantial sources of supply this was one of the most important, and no doubt was always carefully looked after."

This was the Huerta de la Cana, the garden of the palm tree - soon to be obliterated by the operations of RTC.

[Today very little remains of the Roman mining operations. However on the "island" of gossan between Lago and Salomon opencasts, a number of workings at two or more levels in the jarosite have been investigated recently (1981-85) - they vary between less than a metre and two metres high, worked by a form of longwall method, with access from more or less square galleries a metre high, and 70 cm or so wide. In one a heap of jarosite was found, and nearby a perfect example of a Roman lamp. These workings are due to be destroyed, but others at the east end of Lago will be preserved. Included in the latter is a part of the Cueva del Lago, with large chambers, galleries up to two metres high , and circular shafts (all blocked to surface) of a metre diameter. There is evidence that firesetting was used to break the rock, and also of use of both large and small hammer gads or chisels. Willies, Lynn 1981 Report on Mining at Rio Tinto Institute of Archaeology, London Huelva Archaeo-metallurgical Project.]

Figure 7 shows the location of many of the Roman workings - it is based on a plan made by RTC circa 1890. Appendix 1 contains information on Roman activity elsewhere in the Iberian Peninsula.

After the Romans left, soon after 400 AD, mining must have declined very rapidly, and yet the knowledge that they recovered gold and silver and the type of ore worked for the metals persisted - this was made clear when Diego Delgado visited Rio Tinto in 1556.

Figure 5 Entrance to Cueva del Lago

From a photograph taken around 1880. The cave was the source of the Tinto river and probably the place where valuable minerals were first discovered by the Ancients.

Figure 6 Roman Shafts Exposed in Early operations of RTC

Above: Timbered shaft approximately 86 cm square.

Below: Round shaft with hand/foot holes

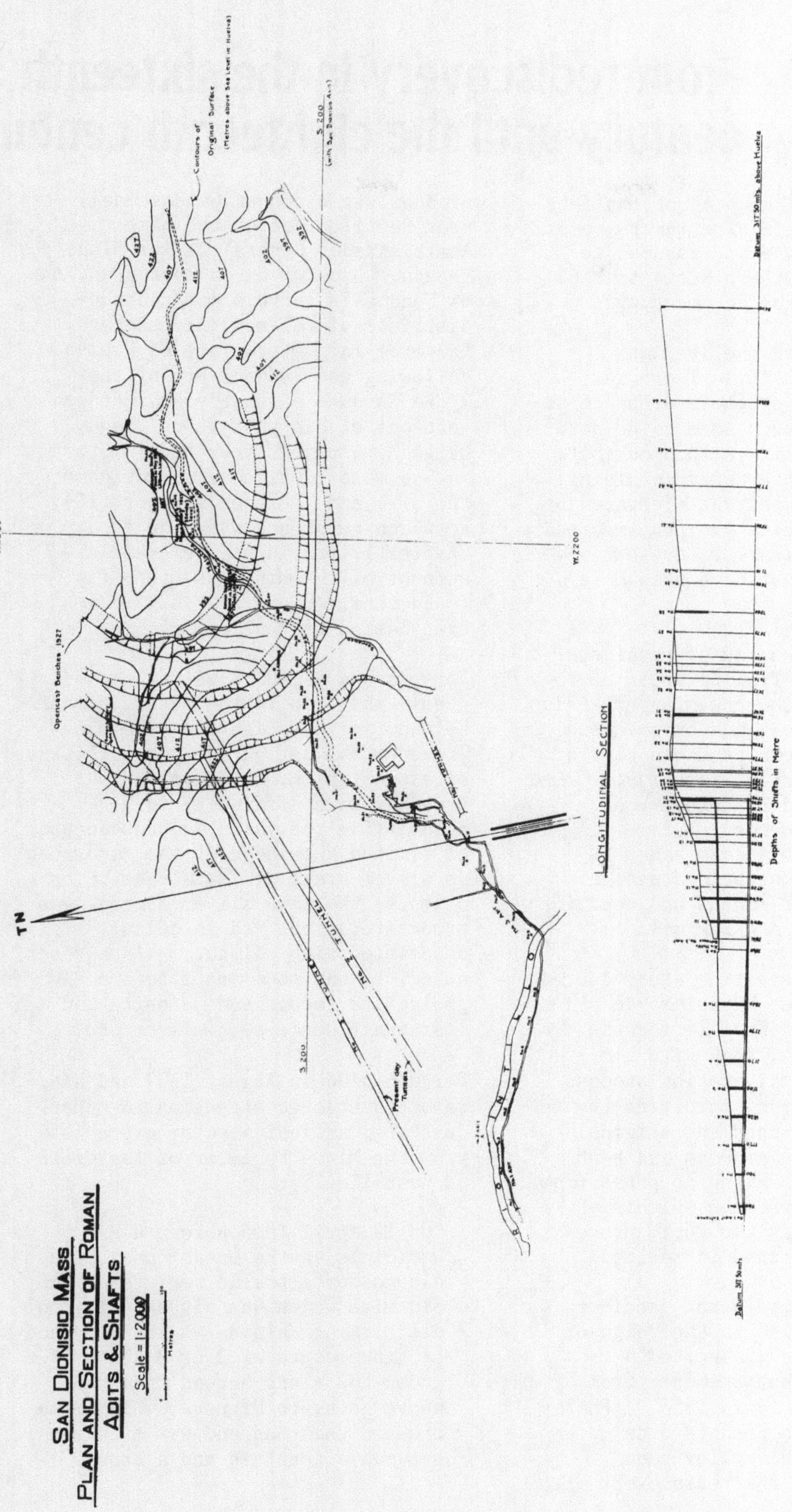
San Dionisio Mass
Plan and Section of Roman
Adits & Shafts
Scale = 1:2000
Metres
Opencast Benches 1927
Contours of Original Surface
(Metres above Sea Level in Huelva)
Present day Tunnels
Longitudinal Section
Depths of Shafts in Metre
TN

Figure 7

CHAPTER 3 From rediscovery in the sixteenth century until the eighteenth century

Nothing much is known about the fate of the mines during the incursion of the Barbarians who are assumed to have reached southern Spain in the second half of the 5th century.

The Moors entered the Iberian Peninsula in 710 AD. There is little evidence of their presence at Rio Tinto, although some coins have been found, and it is assumed that they were not interested in mining. Perhaps some copper was recovered by them by cementation on iron immersed in the acid drainage waters but any such activity must have been on a small scale.

Following the success of Ferdinand and Isabella in freeing Spain of the Moors in 1492, there was an awakening of interest in mining for precious metals. It seems probable that at that time any activities at Rio Tinto had been related to recovery of copperas - ferrous sulphate - deposited on the stream banks by evaporation of the acid drainage waters in summer months and used in dyeing, tanning and ink making.

There is an interesting story quoted by Percy(24) that Bartolome de Medina, who left Seville for the Americas in 1554, took with him details of an amalgamation process for silver recovery involving the use of copperas and that the original discovery of the process had been made by a German using copperas from Rio Tinto. A version of this process, known as the Patio process, was widely practised in Mexico.

The search for mines for precious metals intensified in the reign of Philip II who had inherited a virtually bankrupt treasury from Charles I of Spain in 1556. Philip commissioned Don Francisco de Mendoza, who had earlier been associated with the rediscovery of rich silver mines at Guadalcanal, near Seville, and at the time Administrador General de las Minas de Espana, to carry out a search and one of Mendoza's surveys covered the districts of Zalamea, Aracena and Valverde in the province of Huelva. Following the rediscovery of the mines in June of that year, Mendoza left one of his party, Don Diego Delgado, a priest from Madrid, to make a more detailed examination of the site and Delgado's report(10), to which reference was made in Chapter 1, provides a remarkable account of the mines as his party found them. Appendix 2 is a translation of that report.

For reasons at which we can now only guess, and despite the enthusiasm of Delgado in investigation, the Treasury did not pursue possibilities of exploitation. There was probably some difficulty in identifying the valuable ore because of limited knowledge of the jarosites as silver ores; at Guadalcanal, for example, the main silver mineral was proustite, light red in colour, associated with galena. As already noted, it was some years before RTC geologists recognised jarosite and its significance as a silver ore.

Delgado died in August 1557 and his report attracted attention in other quarters, as indicated by extracts from the Mines Register of the Crown of Castille(25):

> "On 19 April 1565 before H M's Officials at the Guadalcanal Mine, Alonso Gomez Adalid registered an old mine which was situated in the district of Zalamea in the Province of Extremedura at 2 or 3 shots of a cross-bow along a road from the above going to Higuera, which mine crossed the road and was situated between a fountain and a cross."

"On 25 February 1567, before the same officials, Alonso Criado, of Aracena, registered in his own name and that of Anton Criado, a mine of gold and silver,lead and tin and any other metal, situated in the district of Zalamea which is called the "Arroya de las Puercas" in a ravine of the charcoal burners."

"On 8 April of the same year, before the same officials, the same Alonso Criado, in the name of Bartolome Criado, registered another mine in the district of Zalamea which is called the "Cabezas Altas" on the left-hand side of the road which goes from Zalamea to Segunderal."

"In Madrid on 20 September 1569, before H M's members of the Accounts Office, Juan de Cabrera registered various ancient mines and shafts of gold, silver and other metals at sites in the district of Zalamea."

"On 13 January 1570, before H M's Officials at the Guadalcanal Mine, Francisco Perez de Canales, for and in the name of Diego Blanco, of Valencia, registered some large and small slag heaps in the district of Zalamea la Real, from the source of the Rio Tinto and the Salitre Cave to what they call the house "de los franceses."

"On the 17th of the same month, before the same Officials, Bartolome Hernandez registered in the name of Juan del Valle a metal mine in the district of Zalamea."

"On 30 June 1575, before Officials of H M's Accounts Office, Gabriel de Lujan, in the name of Bartolome Sanchez Gigueno, Diego Ruiz, Martin Vazquez, Vasco Yanez, Lorenzo Vazquez and partners, registered a mine of some metals in the district of Zalamea, called "Canada de los Bonigos."

Despite this evidence of interest, no serious operations seem to have been undertaken until 1627 when, on 1 November, King Philip IV directed that the Rio Tinto mines should again be inspected and appointed Don Gregorio Lopez Madera, a member of the Mining Council, and Captain Tomas de Cardona to make a general survey of the Sierra de Zalamea and to "study a certain metal found in great quantities and containing some silver".

No report by Madera has been found but some views put forward by Cardona are quoted by Rua Figueroa(14). According to Cardona, the "metal blanquillo", as he called it, of Zalamea was actually a new metal and he considered that it could replace tin, which at that time was brought from Cornwall, for making bronze castings.

Cardona gave an analysis of the "metal blanquillo" - about 2,800 grams of silver per tonne but little copper - and said it was produced by the Ancients when recovering silver. The Royal Council, to which Cardona's findings were reported, decided that the "metal blanquillo" was a slag of no value, despite the silver content which would be difficult to extract.

It is possible that Cardona's sample was of speiss which is likely to be formed in smelting ores containing arsenic and antimony; substantial quantities of speiss are associated with the slag heaps at Rio Tinto.

In 1637 a certificate was granted by the King to Captain Francisco Moreno de Busto to "go to Rio Tinto and work some mines and galleries in that district in the province of Sevilla." [Provincial boundaries were probably not clearly delineated in this period since Rio Tinto is described variously as in the province of Sevilla and in Extremadura, and is today in the province of Huelva. Ed.] The mines were described as situated on a hill on which there was an old castle called Salomon... with a cave to the west called Murcielago and were found full of water.

In 1661 a certificate was granted to

Don Alvaro Alonso Grafias to use the source of the Tinto river and the slag and metal blanquillo of Zalamea la Real and the hamlet of Rio Tinto to collect copperas and to convert iron thrown into that water to copper - the first reference to recovery of copper by cementation although reports suggest that recovery by this method was not pursued.

In 1693 a certificate was granted to Manuel Fernandez Valle "to continue the exploration of the two mines of copper and silver in the boundary of Zalamea and the pasture land of Malparido, that of silver runs in north to south direction and is called San Benito, the other of copper with some silver is called Senora del Rosario."

The efforts of these, and other entrepreneurs and prospectors of the day, show an increasing interest being taken in the Rio Tinto area and its mineral wealth. The official reports shed some interesting light on their activities; unfortunately it seems that the knowledge accumulated in this period was either overlooked or ignored by those eventually responsible for reopening the mines.

It was almost two centuries after the rediscovery of the mines before a decisive step to rework them was taken. In 1725 a licence was issued to a Liebert Wolters Sjohjelm, probably of Swedish origin, to work five mines in the region for a period of thirty years; the mines were Guadalcanal, Cazalla, Galaroza, Aracena and Rio Tinto and a company was formed to work them for gold and silver.

A mining expert, Robert Shee, probably an Englishman, was called in to inspect the old mines and advise on the working of them and his report has been quoted, in translation from the Spanish, by William Nash(23). His translation is given in Appendix 3.

This peculiar and rambling report has been said to be one of the earliest modern "mining expert's" reports but is not as accurate in many respects as that of the priest Diego Delgado. Delgado was not an "expert" but he found, or was led to, the silver ore, took samples and had them assayed - Shee's report gives no indication of his taking samples and he seems to have relied heavily on the views of another "expert" who examined the old mining activities between 1648 and 1651.

Archaeologically, Shee's observations are interesting. He saw the foundations of a rather large town supposed to have been the Celto-Iberic city of Gran Bitania, sometimes called Betulia, and later a Roman town. The foundations were lost some time after Shee's visit and rediscovered by Luzon(4) in 1967. Since 1967 the area has been covered with a stockpile of gossan awaiting treatment for gold and silver extraction.

Liebert Wolters' company had a turbulent history, described in detail by Avery(1), and following the receipt of Shee's report it was dissolved and two new companies were formed - one, the Compania Espanola, in which Wolters had no interest, taking over the mines of Guadalcanal, Cazalla and Galaroza, and the Compania de las Minas de Rio Tinto y Aracena with Wolters as the only shareholder. On 26 July 1727, only 20 days after his new company was formed, Wolters was killed in a mine accident at Rio Tinto and his rights in the new company passed to his nephew, Samuel Tiquet Sjohjelm.

Rua Figueroa(15), writing in 1859, mentions that the place where Wolters was buried was still identified as the "Hoya de Liberto" - the grave of Liberto; Wolters would have been known by a Spanish equivalent of Liebert as Don Liberto. He was 62 years of age when he died and might well be considered the father of Rio Tinto as he initiated the reworking of the mines and that working has continued uninterrupted until the present day. He died thinking that Rio Tinto was a rich gold and silver deposit.

Also in 1727, the Compania Espanola entered into a contract with Lady Maria Herbert of Powis to drain the Guadalcanal mine, work which was completed in June 1732. Following the failure of the Company to pay for the work, Lady Herbert sued and in March 1740 obtained control of the Compania Espanola. In the following month she laid claim to the mines run by the Compania de las Minas de Rio Tinto y Aracena, on the grounds that no work had been done at either of the mines. Her claim was upheld and, in June 1742, by royal licence, she obtained control of them.

Tiquet appealed against the decree and his appeal was rejected in March 1743. After further appeals and long and costly representations, Tiquet succeeded in getting the decree annulled in July 1746. He was then granted an extension of the lease, for a total of 30 years from that date. Lady Herbert had exploited the mines at Rio Tinto during the four years interregnum and abandoned the workings in a sorry state. Tiquet was forced to raise money to return the mines to working condition by issuing shares - among the new shareholders was a tailor from Valencia, Francisco Tomas Sanz who became a close associate of Tiquet.

Tiquet was the first to recognise the potential of Rio Tinto as a copper mine. In 1737 he had started experimenting with copper recovery from the drainage waters by cementation with iron and had recovered 300 kilos; another 300 kilos were recovered in the period of Lady Herbert's control. When Tiquet regained control, he concentrated on copper production. Initially he used scrap iron; later, about 1749, he introduced virgin iron from producers in Viscaya in northern Spain. Copper production increased from a few hundred kilos in 1747 to 35 tonnes in 1756 - of some 146 tonnes produced under Tiquet's management between 1747 and 1758, only about 5 tonnes was produced by smelting ores.

Rua Figueroa(14) has provided a wealth of information about operations at Rio Tinto from the Wolters era onwards. He found an old report showing the iron usage for cementation over an 11 day period, as follows:

9 May 1752 at midday Viscaya ingots	273	kilos of
9 May at midday scrap iron	77	" "
10 May at 11 am Viscaya ingots	26	" "
15 May at 7 pm Viscaya ingots	557	" "
16 May at 10 am Viscaya ingots	40	" "
19 May at 10 am Viscaya ingots	272	" "

That consumption of 1245 kilos could, theoretically, have produced a roughly similar weight of copper but, since the total copper production for 1752 was only 6 tonnes, it is a safe assumption that iron consumption was several times the theoretical, due probably to high acid and ferric iron concentrations in dilute solutions. The cost of iron is known to have been a major burden on the operations.

The cement copper was melted in a German blast hearth or refining hearth furnace, known at Rio Tinto as "Horno de Capel". It consisted of a hemispherical bowl with a tuyere projecting over the rim at an inclined angle, so that the blast struck the burning fuel and melted copper. A furnace was capable of melting 100 kilos of copper at a time, producing black copper of about 95% Cu. This metal was sent to either the Royal Mint or the Armaments Factory, both of which were in Seville.

In 1750, Tiquet erected a small blast furnace on a piece of ground that had been levelled near the river bank below Lago and used a water wheel to operate the bellows. The site was known as Nuestra Senora de los Desamparados - Our Lady of the Forsaken - and it was here that the smelting of copper ores was gradually developed, in combination with the melting of the cement copper. Ore for smelting was probably mined from the area later developed by RTC as South Lode.

Some 15 years later, after Sanz had inherited management, four further flat areas were prepared in terraces so that the water could be used in descending stages. This site came to be known as "Los Planes" and the four smelters, each consisting of one furnace, were known as San Gabriel, San Jose, San Francisco del Paula and Nuestra Senora del Rosario. An extraordinary thing about the Planes site was that it was 2,500 metres from the mouth of the mine exit, the adit San Roque, where the drainage water was treated by cementation.

Each furnace would have been no more than 1.5 metres in height, square in plan with sides of about 60 cm. The design would have been little changed from Roman times, although the Romans smelted chalcocite and Tiquet had only a small proportion of chalcocite in a charge which consisted mainly of cupreous pyrites. Germans were brought from Mansfeld to operate the furnaces but they found smelting of the Rio Tinto ores more difficult than smelting of German ores. Whereas the German ores required roasting only once to give matte of sufficiently high copper content, Rio Tinto ores needed three roasting treatments and still produced a matte of lower grade which required roasting and smelting a number of times in order to make black copper.

Tiquet cleaned out the Roman adits, first San Roque and later San Luis, as the basis for his mining operations for copper ore. He had difficulty in finding ore of sufficiently high grade, a problem not solved in his lifetime. When he died in September 1758 the work force at Rio Tinto was a mere 14 men and it is remarkable the tenacity which had been shown under his management, with very limited resources in an extremely isolated situation.

Francisco Sanz succeeded Tiquet who had nominated him trustee and administrator. He had been well trained by Tiquet in mining, smelting and administration; he also had natural cunning. The company was in a disastrous financial condition when he assumed control in 1758; when the lease reverted to the State in 1776, all the debts had been paid off, the work force expanded to more than 700 and a substantial community created; the production of black copper increased, averaging 67 tonnes pa over the period.

Reference has already been made to the development of smelting of ore on the Planes site under Tiquet. In 1765 Sanz drove a new adit, at the east end of the lode and at a level 51.7 metres above the Roman adit of San Luis, and named it Santa Barbara - this adit was joined to a shaft of the same name up which ore was raised by hand winch and taken to the Planes area for roasting and smelting.

Sanz made an interesting contribution to the archaeology of Rio Tinto when, in continuing the clearing of the San Luis adit begun by Tiquet, at 110 metres from its entrance, he uncovered a copper plate dedicated to the Emperor Nerva and fixing its date at about 97 AD. Later the area to the east of the mine was named Nerva, and remains so today.

After the Wolters/Tiquet lease expired in 1776, Sanz was retained as the State's Administrator, a title which thereafter was the official one for the Government appointed manager, until August 1783 and for those further 5 years average copper production rose further to 92 tonnes pa.

The relative prosperity at this time brought social changes, most importantly in housing and amenities for the work force. Sanz developed a site near the mine on the slopes of Cerro Salomon which became the aldea or hamlet of Rio Tinto; also referred to as pueblo la Mina. He built 40 houses, a chapel (the Hermita de San Roque), stables, a shop and a tavern. Communal ovens were provided for bread making, fountains were installed for drinking water, market gardens developed and reafforestation started.

Two Franciscan brothers from Granada, Father Gabriel and Brother Pedro Mohedanos, visited the mines in 1780 and wrote glowingly of what they saw:

> "In the neighbourhood of Rio Tinto some evergreen oak trees are found, but the greater part of the locality is covered with a dense growth of jara and other useless scrub, not a single pine, but D. Francisco Sanz, principal Administrator of the Mines, obtained a supply of seeds from Niebla and sowed them broadcast. Today, you can see the development of a fine pine grove from which, some day, the wood will be used for construction and for the miners. He has also planted an area with vines and fruit trees that grow well, serving for recreation and producing fine fruit. These efforts at cultivation we have witnessed with much satisfaction and we understand from several people knowledgeable in this matter that the copper obtained from this mine is good quality and not much blended with iron, as has been falsely stated with notable prejudice to the material and best interest of Spain."

There is no doubt that during Sanz's management the mines were brought into good working order, although he feathered his own nest and practised nepotism on a rather large scale. Many slanderous statements were made against him, especially about private sales of copper and concealing copper stocks. There may have been some truth in statements about the latter - in 1852 black copper was found in the ventilating shafts of the San Luis adit.

After successfully withstanding his critics and opponents for a number of years, starting with opposition to his appointment in 1776 as the Government's Administrator, Sanz was dismissed in August 1783 and thus, at the end of the 18th century, a new era began with the mines operated directly by Government.

CHAPTER 4 The mines under Government control

After the dismissal of Sanz, D Manuel de Aguirre y Horcasta was appointed Government Administrator at the end of 1784 or early in 1785. His first period of service at the mines was short, because he apparently lacked tact and few could work effectively with him. The Director General of Mines, D Francisco Angulo, appointed D Melchor Jiminez to replace Aguirre in December 1786 and he, although respected and presumably effective, had to leave in 1789 suffering from malaria. Aguirre and Jiminez each had a second term and the succession of administrators was as follows:

Manuel de Aguirre	1785 - 1786
Melchor Jiminez	1786 - 1789
Manuel de Aguirre	1789 - 1793
Melchor Jiminez	1793 - 1796
Vicente Letona	1798 - 1825
Jose Martinez Marcos	1825 - 1829

From 1829 the mines were leased to the Marques de Remisa until 1849, thereafter Government again controlled operations directly.

As mentioned in Chapter 3, the community working the mines from the time of Wolters and Tiquet was housed on the slopes of Cerro Salomon in the pueblo la Mina. The village of Rio Tinto, with the parish church and the registrar of births, marriages and deaths, was 3 kilometres distant. Sanz had built a chapel at the hamlet at the mine, in 1786 Aguirre obtained State permission to appoint two chaplains to attend to the spiritual needs of the mine community and in 1789 land was given to the hamlet for a church, the church of Santa Barbara, which was consecrated in 1793 and became the parish church. [This church was demolished in 1918 to accomodate the South Lode opencast and a new church of Santa Barbara was built in the new village of El Valle.]

Aguirre had little expertise in smelting but he tried to smelt the fine calcined ore which at that time was discarded. In an experiment in 1786 he obtained 46 kilos of black copper from smelting 1,400 kilos of calcined ore but the furnace was severely damaged in the process. This confirms a lack of progress since Roberto Shee's report of 1727, which included the following footnote: "Various attempts have been made to smelt small size mineral at Rio Tinto, but without profit, whether the mineral was roasted or unroasted."

When Melchor Jiminez became administrator in 1786, the Director of Mines, Francisco Angulo, with two assistants, visited Rio Tinto and issued a most comprehensive report in July 1788, dealing with all the technical improvements required and recommending means for improving the social welfare of the employees.

Angulo found that the buildings and furnaces were badly in need of repair and that the adits and water channels required cleaning, so that they could be used for cementation. He considered that the roasting of the mineral was being done without sufficient thought of economy and noted that large quantities of roasted mineral were left in heaps. He apparently noticed that copper was being leached out of these heaps by rain and this led him to consider whether roasted ore could be washed by the mine drainage waters, or other water from some other source, to dissolve out the copper.

He proposed to concentrate the solution by natural evaporation until copper and iron sulphates (vitriols) crystallised out. The vitriols were to be redissolved and copper recovered by cementation with iron. His was the first recorded suggestion that artificial cementation, as this process came to be known, should be used at Rio Tinto, but over 50 years were to pass before it was used.

Among other recommendations he made was one that horse whims should be used instead of hand winches to raise ore from the mine, and another that copper should be refined rather than sold as black copper.

A remarkably far sighted recommendation concerning employee welfare was for the creation of a fund to help the sick and infirm. He proposed that contributions should be collected from the workmen according to their wages on the following scale:

1. Those earning from one to two-and-a-quarter reales a day to pay a quarter of a real a week and when sick to receive a real a day.

2. Those earning from two-and-a-half to four reales a day to pay half a real a week and when sick to receive one-and-a-half reales a day.

3. Those earning four-and-a-half reales a day and over to pay a real a week and receive three reales a day when sick.

[A real, one quarter of a peseta, was probably worth 2 or 3 pennies in contemporary English money.]

Very little notice seems to have been taken of Angulo's report by the authorities in Madrid, but the furnaces and canals were repaired and copper refining was introduced. About the time of his report, a new smelter, Los Chaparrales, consisting of one blast furnace, was built on the Tinto river, not far from the village then called Rio Tinto (subsequently Nerva).

Between 1788 and 1792, the average annual production of black copper was 185 tonnes, average copper content 94.2%. Refined copper production averaged 132 tonnes pa so that recovery in refining was around 76%. The refining slag would no doubt have been returned to the blast furnaces and the copper recovered from it included in the 185 tonnes. Refined copper obtained from cemented copper precipitate was 11 tonnes pa.

Villagers from both Zalamea and El Campillo had regularly encroached on the mining lease with their pigs, taking timber, and once were responsible for burning a large part of the woodland. When Aguirre returned as Administrator in 1789 he went to some trouble to establish the lease boundaries - with full cooperation from representatives of the local interests, Andres de Canete, a surveyor from Seville, was appointed to conduct the survey and he produced a plan acceptable to all parties in August 1791. Rua Figueroa wrote that a copy of the plan was held in the town hall of Zalamea when he worked at Rio Tinto in the mid-19th century but the two copies given to Rio Tinto could not be located.

Vicente Letona was appointed administrator in November 1798. He had been employed at the mines for many years, having arrived there in Sanz's time and had married his granddaughter. His appointment covered 27 years and spanned the difficult period of the Peninsular Wars, during which the country was frequently overrun by the French. During such times he had great difficulty in maintaining production and all too frequently Madrid could not supply funds - in such extremes he used his own on many occasions.

Copper production from 1793 until 1800 averaged 15.5 tonnes pa; the difficulties created by the War are reflected in the fall to an average of 8.2 tonnes for the years 1801 to 1809. Production of copper precipitate stopped in 1810 and the smelting furnaces ceased operation in 1811. Seville was captured by French troops in 1810 and their commanding general, Jean Pierre Maransin, requisitioned all available copper for munitions. At considerable risk to all involved, most of the copper at the mines was sent to Cadiz, for some time the only bastion holding out against the French. It was 1814 before any further production of copper precipitate and a further 10 years before smelting restarted in 1824.

Throughout this period, the population at the mines suffered severe hardship and many, including the skilled workers, left to seek work elsewhere. By 1812 the mines and works were in a ruinous condition and rumours reaching Madrid suggested that Letona was responsible. In January 1813, Domingo Ibarrola, a Government accountant based at Algeciras, was sent to investigate - his report gave credit to Letona for his efforts under difficulties and recommended further support for him.

A technical manager, Jose Miaja y Pingarron, was appointed with an assistant, Jose Martinez Marcos, but alas no financial aid was forthcoming. Jose Martinez made changes to the cementation canals and started using scrap iron instead of new metal, although even scrap was difficult to obtain. Production of precipitate copper improved but was erratic - 27 tonnes in 1820, 12 tonnes in 1823. Stalactites formed in the mine , containing 4 to 12 % Cu and 2 to 10 % Zn, were first used at this time as vitriol, dissolved in water and the copper recovered by cementation, a practice which was to continue until 1893.

Despite improvements made by Jose Martinez, recovery of copper from acid waters was not good and, in 1822, D Juan Santana Bolanas was granted permission to treat the waste water from the cementation canals. Between 1824 and 1827 he produced about 19 tonnes of copper and paid the administration 400,000 reales, say £4,000. [The economics of this operation are intriguing - according to Avery, copper was worth just over £100 per ton in Spain about that time so it appears to have been a good deal for the administration - Ed.]

Jose Martinez succeeded Letona as administrator in 1825 and about this time the Planes smelting furnaces were rebuilt closer to the village of Rio Tinto, alongside the Rio Agrio, upstream of its junction with the Tinto river. Water flow on the Rio Agrio was not sufficient to drive the bellows for the smelters through the year - at times manpower was used - and the feasibility of bringing water from Cueva del Lago was examined but not pursued.

That feasibility study was based on a detailed survey made of the mines area in 1828 by Joaquin Esquerra, and still in use in the 1920s; Esquerra's survey had been commissioned by D Fausto de Elhuyar de Zubice, Director of Mines, who had visited the mines in 1823. At that time of deep depression, he believed that government management could be successful if funds were provided, and expressed confidence in finding a means of recovering copper from the "small size mineral found in the mine". Because of lack of funds, government did nothing to restore the mines; instead it was decided to lease them to private enterprise and it was for that reason that the survey of 1828 was commissioned.

The 45 years of state control covered some of the most difficult years in the country's history, with the Peninsular Wars bringing widespread economic ruin and this aspect of the period is covered in some detail by Avery(1).

CHAPTER 5 Private control, 1829-49, the Marquis de Remisa

Following the advertising of the lease in April 1828, three offers were received by the end of August, the closing date. The offer accepted was made by the Marquis of Remisa and he obtained a lease of 20 years from 24 April 1829 at a rental of 67,000 pesetas pa for the first 10 years and 125,000 ptas for the next 10, say £3,000 and £6,000 pa respectively.

The lease carried a number of conditions which, in the light of what followed, might be better described as "best endeavours" directions. These were based on recommmendations made by Fausto de Elhuyar de Zubice, Director of Mines, in May 1829. Remisa was to install a new shaft with hoist, wheelbarrows were to be used to convey ore to the hoisting shaft to replace the practice of moving it along galleries in wooden containers passed from man to man. He was also to improve the drainage of acid waters from the San Roque adit to facilitate treatment for copper recovery.

Remisa did only what he considered important; since he appears to have had influential friends in Madrid no sanctions were used against him. Although the new shaft was not sunk, the old Santa Ana shaft was improved and fitted with a whim.

A retired army officer, D Alejandro Viviente Espeleta, was appointed by Remisa as administrator of the mines and he introduced his own ideas. For example, assuming costs would be reduced by roasting ore and smelting the resulting calcines as close as possible to the Santa Ana shaft, he built two blast furnaces there. It cannot have been only the lack of water power to operate the bellows which led to the failure of this project after 3 years of working - bellows were commonly worked manually at the other smelter sites. Since the Santa Ana shaft was quite close to the pueblo la Mina, it may have been persistent objections to the sulphurous roasting gases which eventually brought the operations to an end.

The scale of operations was substantially increased, notably in the heap roasting of ore for smelting - some details were reported(26) by the Director of Mines, D Luis de la Escosura who visited Rio Tinto in 1844 or 1845. Initially roasting was done in conical heaps about 5 metres in diameter and 2 metres high containing about 44 tonnes of ore. Larger, rectalinear heaps with rounded ends were developed and were proved more economical of fuel - these heaps were known as "teleras" because their shape was similar to a loaf of bread of that name made locally. The fuel used was wood - the local oak - and vast quantities were required. Escosura gave the following figures:

	Conical Heaps	Teleras
Pyrites, kilos	5,000	5,000
Wood, kilos	490	390
Labour, reales	9.0	6.5
Roast time, hours	110	?

Single roasting achieved only partial removal of sulphur and three roasts in all were needed to produce calcine for blast furnace smelting, as explained in Chapter 3. The three roasts took 20, 8-10 and 8-10 days respectively for the conical heaps and 30, 20 and 15 days for the teleras. Like his predecessors, Remisa used blast furnaces to smelt calcines to black copper but he introduced reverberatory furnaces for refining the black copper.
Details of the blast furnaces used are shown in Figure 8. The furnaces

were built in pairs, enclosed in a strong stone built wall on three sides and with linings of refractory slate. The front wall was removable, with an arch above to help support the chimney. The front, 35 cms thick, was also made of slate, with separate tap holes for slag and copper. The charging port was about 1.7 metres above floor level, probably a convenient height at which to discharge baskets or wooden trays carried on the heads of the workmen. Hearth dimensions were 50 cms by 83 cms and the tuyere entered through the back wall at a height of about 40 cms above the hearth - the hearth itself was made of clay mixed with charcoal and appears to have been rather easily eroded by slag.

The furnaces were known locally as "Alemanes", a name probably dating back to the period after 1750 when German smelting specialists from Mansfeld were brought to Rio Tinto by Tiquet. The drawings in Figures 8 and 9 are taken from Escosura's paper of 1845.

A daily smelting charge to a furnace consisted of :

Calcine	1140	kilos
Ancient slag	1600	"
Charcoal	1370	"

This charge produced 70 to 80 kilos of black copper. Ancient slag was the only source of silica used and Escosura refers to sows of iron produced in the furnace because of the lack of sufficient silica to slag all of it. Furnace campaigns were of about 5 days - the front wall was then taken down and the hearth remade.

The reverberatory or "spleiss" furnace used for refining black copper was also of German design and details are shown in Figure 9. The round hearth was about 1.7 metres in diameter, with a tuyere at "c" connected to a bellows. Wood was the fuel; flue gases escaped through the semi-spherical roof at ports marked "x".

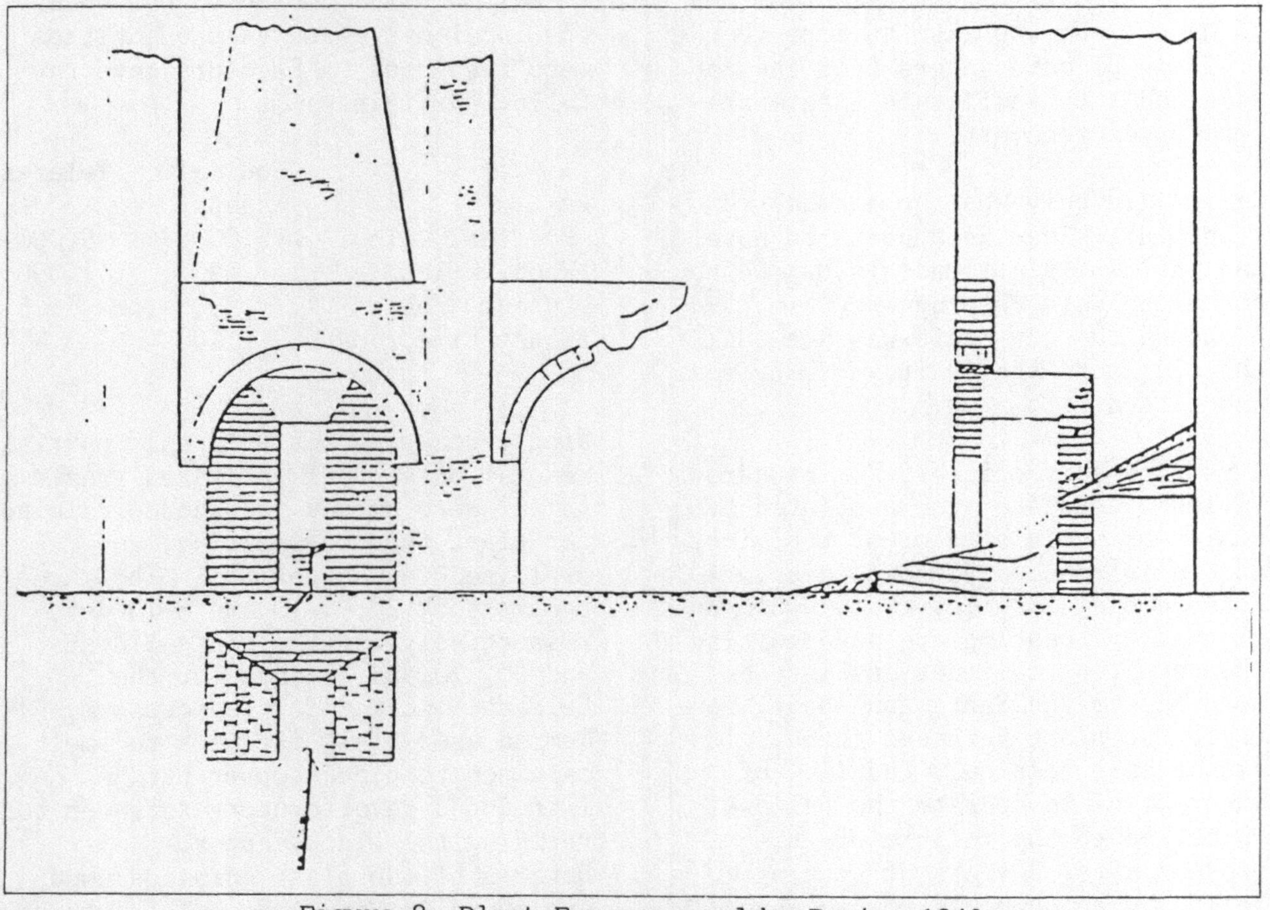

Figure 8 Blast Furnace used by Remisa 1840s

Black and cement copper were melted slowly to oxidise arsenic and sulphur. Once the copper was melted, the temperature was raised, slag removed and air blown on to the surface to oxidise some of the copper, at the same time oxidising further arsenic, iron and other impurities. The temperature of the bath was lowered to remove dissolved sulphur dioxide, slag was again removed, and the surface was covered with charcoal to reduce the copper oxide.

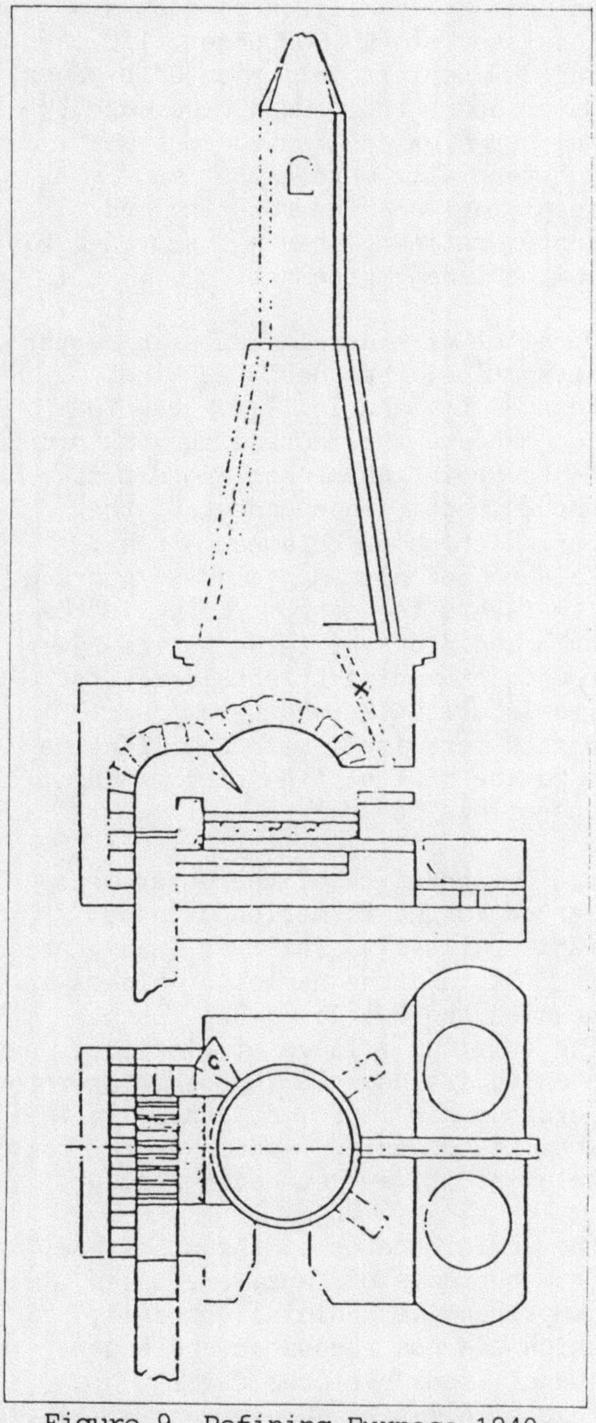

Figure 9 Refining Furnace 1840s

When a small spoon sample showed a rose-red colour the copper was tapped and cast into rosetting pots; the surface was cooled by water and thin circles of copper, rosettes, removed. Refining slags were resmelted from time to time to black copper.

This method of refining was used up to the time of the RTC and was still used at a much later date in Germany and Hungary. However, Escosura was not satisfied with the quality of refined copper produced at Rio Tinto; at that time it may have contained residual arsenic and was further refined at the armaments works in Seville.

The major technical developments of the Remisa era were associated with the production of copper by cementation. In 1839 Remisa formed a separate company, operating at Planes under the direction of D Vicente Lopez Probe, to extract copper from crystalline vitriol, the most important sources of which were stalactites and "tierras vitriolicas", small sized, oxidised mineral also recovered from mine workings. The vitriol and "tierras vitriolicas" were leached, or "washed" in the local terminology, and the copper cemented from the resulting solution by iron.

Copper produced by cementation in this way was more pure than that obtained by cementation from mine drainage water and was separately refined; the latter, known as "cascaria de la caneria" and containing significant arsenic, was refined in the reverberatory furnace with black copper.

Around 1841, a D Felipe Prieto visited the mines and saw the roasting of the cupreous pyrites for the smelter - he claimed he knew a process which could recover copper from the low copper content pyrites then being discarded. Remisa gave him the opportunity to prove this process, generally known as "artificial cementation"; it consists of roasting at low temperature, which converted copper sulphides to water soluble sulphates, and leaching the copper from the roasted mineral. The roasting, called "slow roasting", involves a higher degree of sulphur elimination compared with the high temperature roasting for matte smelting.

Acid water from the Cueva del Lago was preferred for the leaching and up to 70% recovery of copper from ores of less than 3% Cu was obtained - with higher grade ore recovery was lower; as little as half that figure. Copper remaining in the calcines or cinders after the leaching operation was slowly solubilised by natural oxidation. The cinders dumps were known as "Barbasco Heaps" after RTC took over, earlier they were known as "terreros", the name used later by RTC for the areas used for natural cementation.

Prieto was granted a patent for his process for 15 years, commencing 9 September 1845 and a new company, the Compania de los Planes, took over the business of the earlier company recovering copper from vitriols and oxidised mineral. Provincial authorities in Huelva objected to the operations of the new company; it had not been registered in Madrid and caused serious damage to State property, destroying trees with sulphurous smoke from the burning heaps. Remisa, with his influential friends in Madrid, persuaded the authorities there to approve the company and its operations and it continued in profitable activity after his lease on the mine itself had expired.

The Prieto patent expired in September 1860; thereafter the process was used by all the significant pyritic mines in the south-west of the Iberian peninsula. However, open air roasting was prohibited in Portugal in 1878 and a similar prohibition was made by the Spanish Cortes (parliament) in 1888 and this greatly curtailed the use of the process.

From 1842 the production of copper from vitriols had increased considerably. Although less ore was smelted, total copper production increased from 72 tonnes in 1829, the first year of Remisa's lease, to 427 tonnes in 1848; the annual figures varied widely and are given in Appendix 4.

Of many conditions of the lease with which Remisa did not comply was one concerning woodlands. When he took over the property the trees were valued at 3.8 million pesetas and consisted of 440,000 pines, 170 oaks and 300 poplars, all planted by Sanz. Quite apart from the damage done by sulphur dioxide, Remisa consumed substantially more timber in supplying wood for roasting and copper refining than was provided for in the lease agreement.

Parallel with developments in copper extraction, the pueblo la Mina, founded by Sanz in 1772, grew to accommodate the increasing workforce, to the point where Remisa felt it should become independent of the jurisdiction of Zalamea la Real. Zalamea had always jealously guarded its rights to control mining in the area and did everything in its power to hinder mining by outsiders, yet its inhabitants, who were agriculturalists, were unwilling to take the risk of financing mining operations themselves.

On 12 February 1841 the Government agreed to the formation of a new municipality. This was inaugurated on 3 May and the hamlet or aldea was renamed the Pueblo of Rio Tinto. The original village of Rio Tinto changed its name to Ventosa, later to Lugar de Rio Tinto, and then, in 1891 when it too gained independence from Zalamea, it was renamed Nerva.

The administrator of the mines had, from the days of Sanz, exercised comprehensive judicial authority which was now renounced and a new council took over the civil responsibilities. Government still maintained that the village was Crown property and, as such, was included in the lease of the mines, even though at that time many of the people living in the Pueblo owned their houses.

Two interesting commentaries relating to the mines during this period are quoted in Appendix 5. The first is by Richard Ford, an Englishman who visited the Rio Tinto area in 1832 or 1833 - even before Prieto's process was introduced the impression was certainly not of a green and pleasant land. The second commentary by Fernando Bernaldez concerns the metallurgical operations as he interpreted them a few years after the termination of the Remisa lease.

During the twenty years of control of the mines by Remisa, considerable, if uneven, progress was made. The operations were undoubtedly profitable and full payment of lease charges was made. Government inspectors were well aware of infringements of lease conditions and the reduction of the woodlands and in the value of the mines installations has been clearly recorded in their reports. Remisa's operations must have been responsible for major damage to the environment, although that was probably not taken into account by officials in Madrid. It is no surprise that on termination of the lease in 1849, his application for an extension was rejected and the mines reverted to government control.

Figure 10 Opencast Operations Exposing Old Workings

19th century pillar and stall workings clearly visible, Roman shafts not positively identifiable in this early 1900s photograph.

CHAPTER 6 The mines returned to Government control, 1849-73

When the Remisa lease expired in 1849, D Casiano de Prada was appointed to take over the administration of the mines. He was no stranger to Rio Tinto, having been there in 1847 on an inspection for the government; he was soon in difficulties, as he later wrote(27):

> "On 24 April 1849 the contract for leasing the Mine terminated.
> On the following day I put into working order and operation all sections of the works as on account and on behalf of the Government, without being able to count upon, or being furnished with, the necessary funds for even the most urgent and important matters, although I did not anticipate that many days would pass without such being sent to me. But I was mistaken and if the labours in the mine and all the works and office did not suffer any interruption that was due only to my determination and to my occasionally procuring funds and loans from Seville, obtained however under my personal guarantee and employing my own private means, sometimes even remaining without any money at all."

He expressed the difficulties and frustrations experienced by many administrators, both before and after his time, caused by the utter and complete lack of understanding by the government officials in Madrid of the problems of operating the mines.

Felipe Prieto signed a new contract on 23 January 1849 which enabled him to continue working until the expiry of his patent in September 1860 - under the contract the mines were to supply 2,300 tonnes of suitable ore per month and he was to return 17 tonnes of copper monthly. By the end of the contract period the mines had supplied 270,000 tonnes of ore, somewhat less than the contracted amount, and Prieto had delivered an average of almost 25 tonnes of copper per month .

Prieto's Planes Company used drainage water from the Cueva del Lago for leaching, and the location of its works, Fabrica de los Planes, is clearly shown on an 1878 RTC plan, a reproduction of which is inserted inside the back cover of this volume. It seems probable that the Company leased the Chaparreles smelter to melt and refine copper precipitate - not far from the site of this smelter the river bank consists of dumped slag which is high in copper, approaching 2%. It is known that Prieto's slags were high enough in copper to justify re-smelting in later government operations.

About 1850, the Treasury made an agreement with D Mariano la Cerda, a priest attached at one time to the church at Rio Tinto and with influential friends in Madrid. He claimed to have an electrochemical process for recovery of copper from liquors obtained by leaching calcined cupreous pyrites, a process vastly superior to other known methods. His claims were received with enthusiasm by the Treasury, although not necessarily by the Minister of Mines.

Without any serious investigation of the claims, Cerda was granted a patent and began exploitation under the agreement entered into with the Treasury. Exploitation required the use of Prieto's slow roasting method; how he got over that difficulty is not known. As far as the electro-chemical label is concerned, it would have been fashionable and impressive at that time since it had been introduced when Michael Faraday published his paper on "Electrochemical decomposition" in 1834.

The Treasury agreed to Cerda taking half the residual mined ore after the obligation to supply the Planes Company had been met and to leasing the Desamparados smelter. It is thought that Cerda took delivery of ore near the Roman adit of San Pedro and, to save transport, roasted it in the same area, very near the village, Rio Tinto pueblo. The location of his Fabrica de Cerda is shown on the 1878 plan.

Cerda's story is told in some detail by David Avery(1); his operations ceased in 1858 when it was recognised that they were similar to those of the Planes Company. His enduring legacy to Rio Tinto is that the area to the south-east where his roasting heaps were located has been known as Cerda for generations.

In 1856, two mining engineers, D Antonio Anciola and D Eloy Cassio, inspected the mines and in their report(13) suggested that opencast mining would be cheaper and safer than the pillar and stall method then employed. That method had been introduced in 1830 and was not successful because of the large number of Roman galleries and caves which made it impossible to leave the pillars at each level one above the other and accidents had occurred due to collapse of galleries.

Anciola and Cassio also suggested that higher copper recoveries would be achieved by smelting copper-rich pyrites; the overall recovery at the time from roasting and leaching seems to have been significantly less than 50% and decreased with increasing copper in the ore. They carried out sampling in the mine and concluded that the average copper content was 4.44%, a figure thought too high by most engineers at the time.

Other engineers also advocated smelting copper rich pyrites but the advice was not put into practice, presumably because of the scarcity and cost of charcoal, and heap roasting followed by leaching and cementation continued.

Several administrators were involved in the running of the mines during this period of State control but real progress was not made until the appointment of Rua Figueroa in January 1859. Although he understood the barriers imposed by Treasury control, he and his assistants made great efforts to increase extraction from the mine and to improve the metallurgy. Output of ore increased in his first year by 23% to 15,700 tonnes and by the time he left in 1863 had reached 89,000 tonnes, an output never again achieved under State control.

Lee Thomas gave an interesting account(16) of the metallurgical operations at the time of his visit in 1862 or 1863:

> "The adoption of the cementation process, as it is now practised, arose necessarily out of the increasing scarcity and dearness of fuel. In the vicinity of Rio Tinto, indeed throughout those parts of the province of Huelva I have visited, a more than unusual dearth of such timber as would be applicable to the manufacture of charcoal is observed. The country is covered with brushwood (gumcistus) of an inferior quality, but one may travel for miles without meeting with an encina tree of any kind. The teleras are open to the air...... The prevalence of wind and rain consequently exercises a considerable influence over the calcination. The exposure to strong currents of air gives rise to fusion of a considerable portion of the sulphurets and the decomposition of the sulphates which pass into oxides and insoluble salts and during the heavy rains the sulphate of copper must be washed away as soon as formed.
>
> "The calcined mineral is carried on the backs of mules or donkeys to the lixiviation tanks. These

are made of rough masonry and lined with asphalt, the common size is 7 metres long by 4.25 metres wide by 1 deep. They are two-thirds filled with calcines and contain about 17 tons."

Of the mine itself, Lee Thomas wrote:

"It contains six levels. The first and second are in a very bad condition. The third, fourth and fifth run into one another in such a way as to make it difficult to say where one begins and the other ends. The sixth, which is on the level with the San Roque adit, has been opened up of late years and the plan of working adopted has been systematically carried out. The dimensions of the levels are 3 varas wide and 2.5 varas high, and the pillars left are 4 varas square. The measurement from level to level is about 9.5 varas and is made up thus:

Height of level	2.5 varas
Underhand stope	4.0 "
Floor left	3.0 "

The method of working necessitates the leaving in the mine about 58% of the ore for floors and pillars."

A vara is 83.5 cm. Thomas is thought to have been mistaken about the levels and there were a total of nine, with only eight which could be drained by the San Luis adit effectively used.

Rua Figueroa(14) has provided a wealth of technical information on leaching and smelting operations in his time. For instance, the calcines were washed seven times and the average copper content of the liquors taken off was:

Wash	Duration	Cu gm/m^3	°Baume
1st	18 hours	18,210	32
2nd	24 "	10,818	15
3rd	30 "	4,971	6
4th	34 "	2,786	3
5th	40 "	1,241	2
6th	40 "	778	-

1 cubic metre of water was required for 1.5 cubic metres of calcines washed.

The wash waters were run into settling tanks, to deposit as much as possible of the suspended matter, and so manipulated to have a final copper content of about 8,000 gm per cubic metre and a density of between 12 and 15 degrees Baume. This liquor went to cementation tanks, of the same dimensions as the leaching tanks and filled with pig iron. Cementation was a batch process and the time taken to precipitate the copper varied from 24 to 30 hours in summer to 40 to 48 hours in winter. In cold weather the liquor needed agitating. The discharge from the cementation tanks contained about 200 gm Cu per cubic metre (ie 200 ppm); it was mixed with natural drainage mine waters, for example from San Luis adit, and directed along a channel containing iron and the final exit water contained about 20 gm Cu per cubic metre. The fine precipitate, papucha, collected from this canal contained about 10% Cu - prior to about 1864 little could have been done with this material because of its high arsenic content.

The cementation or precipitating tanks were cleaned out after having been filled three or four times, when the coating of copper on the pig iron was thick enough to be removed by hand or pick. The tanks were then emptied, all the pig iron with copper attached removed and the fine precipitate on the tank bottoms was shovelled out. Each unit of copper precipitated consumed about 2.2 units of pig iron.

Rua Figueroa gives analyses of two precipitates; the first removed from pig iron, the second collected as sludge from cementation tank bottoms, as follows:

Cu	65.47%	29.72%
As	2.48%	2.77%
Sb	0.64%	1.13%
Metallic Fe	3.64%	ND
S	ND	2.14%

Graphite C	2.15%	ND
Insoluble	6.25%	7.66%

ND = not determined

The precipitate was made into balls of about 10 cm diameter and dried, either in the sun or on a hot floor heated by brushwood, then calcined in a vertical brick furnace fired with brushwood. The depth of charge in this furnace was about 1.5 metres, the brushwood burned for about 5-6 hours, by which time the charge was hot enough to continue calcining autogenously for 6-8 days. Weight loss on calcination was 28-32% for precipitate from artificial cementation and 34-40% for precipitate from natural cementation. Calcination was a very unpleasant operation because of partial removal of sulphur, arsenic and antimony in fume.

Calcined precipitate was melted in a refinery hearth of the type used since the days of Sanz and the process required large amounts of charcoal. Producing black copper, 2,500 kg of precipitate yielded 1600 kg in 24 hours for a consumption of 640 kg of charcoal; producing refined copper the output in 24 hours was between 400 and 620 kg, depending on whether it was copper for alloying, containing cupreous oxide and known as "punto de martinete", or refined copper for rolling. Charcoal consumption in refining was 500 and 620 kg respectively.

Black copper was normally refined in the reverberatory "spleiss" furnace installed by Remisa in 1831. The charge was 4,400 kg of black copper and 1,000 kilos of copper recovered from melting or smelting refinery slags and ladle skulls. Recovery into refined copper was 88%, the melting and refining time 18 to 20 hours and fuel used was 4,200 kg of logs.

Slags from the refinery hearth and the reverberatory furnace were mixed with roasted matte, papucha or other high-arsenic precipitate and roasted kernels, the nature and source of which are explained below. The mixture was smelted with Roman slag in a small blast furnace, with charge and output as follows:

Roasted mixture & slags	1025 kg
Roman slag	1360 "
Charcoal	600 "
Black copper produced	227 "

Slag contained about 0.7% Cu

Operations of smelting precipitate and refining copper were labour intensive by modern standards. The crew for the refining hearth was 5 men per shift of 12 hours; 2 worked the bellows. The reverberatory employed 6 men per shift, although at the melting stage the demands on them were light.

It was Lee Thomas who noticed that kernels formed during roasting in the teleras were not collected separately and used a valuable source of copper to the blast furnaces and he pointed this out to Rua Figueroa. These kernels, called "nucleos" at Rio Tinto, are formed in the slow roasting of cupreous pyrites by preferential oxidation at the particle surface of iron sulphide - cupreous sulphide diffuses slowly to the centre to give concentrations as high as 40% CuS in a central core, encased in a soft shell of iron oxide which is easily detached.

The artificial cementation process produced copper relatively cheaply from low copper content pyrites. There were a number of disadvantages to its use at Rio Tinto and Rua Figueroa listed them as follows:

A considerable amount of fuel wood was required.

All the sulphur in the ore was lost.

It was difficult to roast teleras uniformly; some ore was over roasted, some not roasted at all.

Soluble copper was lost during the

rainy season.

Sulphur dioxide emissions killed the vegetation.

Rua Figueroa observed that pyrites lying on the floor of the mine released soluble copper, the basis for natural cementation, but did not pursue experiments to test such possibilities.

In this period, smelting was still applied to calcined high grade ore but on a reduced scale. By 1864 there was only one small blast furnace working - it was 90 cm long by 60 cm wide and 270 cm high and it had a hearth 60 cm deep made of the usual mixture of fine charcoal and clay. It had one 4 cm diameter tuyere placed 60 cm above the hearth, with blast supplied by a bellows worked by 2 men when water power was not available. Figures from 1841, in Remisa's time, show that the furnace was then smelting 1,350 kg of calcined ore per 24 hours, using 29% charcoal in the charge and producing 67 kg of black copper containing about 92% Cu and probably performance had changed little over 2 decades.

At last, in 1864, an effort was made to act on the recommendations made by many engineers over the years, and notably by Anciola and Cassio in 1856, to develop matte smelting for higher grade cupreous ore. The old Los Desamparados smelter, which had been used by Cerda until 1858, was chosen as the site for another smelter and two furnaces were erected, one 100 cm long by 60cm wide and the other 80 cm by 56 cm, internal measurements; the height of both was 350 cm. Each had a single tuyere of 13 mm diameter placed 250 cm below the top and the hearth tamped from clay and charcoal had a depth of 50 cm. The furnace walls were of porphyry lined with firebrick. Two tapholes were provided for slag and one for matte. A new waterwheel was built, with an output of about 2 HP to drive bellows able to deliver 4 cubic metres at 15-19 cm water gauge.

The furnaces were blown in on 9 January 1865 and, after experiencing some of the difficulties common to such start-ups, demonstrated an improved performance. Campaigns lasted up to 10 days, double that of the earlier furnaces, and the two furnaces were capable of smelting a total of 14 tonnes of charge per day, comprising:

29.0%	calcined ore of	7.2% Cu
10.2%	crude ore	6.0% "
14.8%	siliceous ore	5.6% "
9.3%	kernels	20.0% "
37.7%	slag	1.6% "

Charcoal consumption was 20% of the charge, compared with 29% on the older furnace. Daily matte production was about 3.5 tonnes containing 20-25% Cu.

Treating low grade matte had always been a problem and it became more serious with this lower grade matte which had to be roasted as many as 8 times. This was done in heaps with all the disadvantages already identified by Rua Figueroa for roasting cupreous pyrites.

At Rio Tinto the base of the matte heap had a diameter of between 4 and 5 metres and for the first two roastings the height was not greater than 30 cm. The heaps were then made cone shaped for subsequent roasts and reached a height of 2 metres before the eighth roast. The heaps were built on a bed of logs and brushwood - for the earlier roasts little fuel was used to avoid melting the matte; as the sulphur was removed more fuel and higher temperatures were used.

Among examples of matte roasting given by Rua Figueroa is one for matte obtained from smelting kernels or papucha involving five stages. Roasting 30 tonnes of such matte consumed about 2.5 tons each of brushwood and logs and the complicated series of operations extended in time over at least two months. Rua Figueroa compared this process unfavourably with that used in South Wales where, at that time, roasting 36 tonnes took about 12 days and the resultant product was lower in sulphur than at Rio Tinto.

Roasted matte was smelted in the same blast furnaces used to smelt calcined cupreous pyrites and the smelting and the resultant black copper was refined in the reverberatory furnace, in much the same way as described above for copper from smelting precipitate.

In 1865 the cost of producing refined copper from cementation was 1.39 ptas/kg, compared with 1.29 from smelting ore. The copper from smelting was also of better quality as a result of more effective elimination of impurities in matte roasting. Comparative analyses were:

	Cementation	Smelting
Cu	99.004%	99.711%
S	0.059"	Tr
Sb	0.297"	0.289"
As	0.640	Tr

Arsenic was an important impurity in copper produced at Rio Tinto, and elsewhere in Spain, and presumably the copper from ore smelting was vastly superior material for the Seville armaments factory's rolling mill.

Production of copper reached a peak of 1,335 tonnes in 1863 in Rua Figueroa's final year in charge of operations and gradually declined thereafter. Details of production for this period of government control are given in Appendix 6.

Transport of copper to Seville by horses and mules over tracks was always troublesome and in 1840 the building of a road was suggested but the scheme did not materialise. In their 1856 report, Anciola and Cassio recommended the construction of a railway to a port; this showed considerable foresight on the part of these two engineers who probably envisaged the use of sulphur in the pyrites for the production of sulphuric acid. From the 1840s brimstone was slowly being replaced by pyrites for this purpose and in 1852 Leblanc manufacturers used about 270,000 tonnes for acid making.

A commission was appointed on 5 January 1860 to investigate the construction of a railway. It recommended that the railway should go to the port of Huelva, at that time only a very small fishing port. Lee Thomas(15) was appointed as advisor and was in Spain in 1862-63. Anciola and Cassio had suggested transport for 470,000 tonnes but, since the total mined in 1863 was only aroud 90,000 tonnes, a figure of 240,000 tonnes pa was adopted.

The railway would bring stores, and fuel collected en route, to the mines - it was planned to run from Rio Tinto to Pozuelo, Valverde, Trigueros and Huelva, a distance of 61 kilometres, and to be constructed as a tramway, meaning that no passengers were to be carried, using 40 lb rails, at an estimated cost of £238,000, as follows:

26 miles railway between Huelva and Valverde at £4,000 per mile	£104,000
12 miles railway between Valverde and Rio Tinto at £7,000 per mile	£ 84,000
Compensation for land not belonging to the Govt.	£ 20,000
Plant, waggons, engines, horses, etc.	£ 30,000
	£238,000

This is an early example of the introduction of British units to Rio Tinto; the lump sum estimate was translated to 23 million pesetas. As with so many recommendations, nothing was done, yet the mining operation by a British company at Tharsis received permission for a railway which was operating by 1866, and in 1867 the Buitron mine was granted a concession to build a railway to San Juan del Puerto, a distance of 40 kilometres. [This latter railway was subsequently extended to Zalamea and proved useful to RTC - it finally closed in 1969.]

Another commission was formed in 1867 under the Inspector of Mines, Salazar, and with Cassio as one of its members. Among its recommendations were:

The San Luis adit and other galleries should be enlarged to install rail transport for ore, with an estimated cost saving of 20 reales per tonne.

Opencast working should be introduced, for an estimated further cost saving of 20 reales per tonne and providing opportunity to increase output.

Smelting of cupreous pyrites should be increased - for better copper recovery, saving in pig iron for cementation and increased copper production.

A large adit or tunnel should be constructed to traverse all the lodes at a level 196 metres below the Santa Ana pithead - this tunnel being capable of taking waggons to be moved directly to Huelva on the projected railway.

It was believed that by making these improvements the cost of copper would be reduced from 5,600 reales per tonne to 4,400 if the first recommendation was implemented, to 3,200 if recommendations 1 and 2 were implemented, and progressively to 2,800 and 2,500 by implementation of the further recommendations. Some work was done on enlarging the San Luis adit and, in 1870, a new shaft, Los Inocentes, was started but lack of funding from the Treasury prevented any significant development.

Production costs for 1869, when 233 teleras were roasting and 822 tonnes of copper were recovered by cementation, were as follows:

	ptas
Mining	7.15
Roasting,cementation and refining	15.02
Administration	1.40
Total,for producing 11.70 kg copper from 1 tonne ore	24.34
Cost per kg Cu	2.08
Per tonne	2,080

At that time the price realised for copper in Seville was 1,400 ptas per tonne and the Treasury concluded that it would not be wise for the State to finance further development; the mines were put up to auction and a new era in control and operation was about to start. As Gonzalo y Tarin(7) wrote a few years later: "the bad luck which always accompanied State management was the reason why the idea of its sale acquired a great many backers, even among the Mining Engineers at Rio Tinto, who came to the conclusion that there was no other way to make the mines viable in the future."

CHAPTER 7

Early mining operations of The Rio Tinto Company Limited

In February 1873 the bid of a consortium of British and German business men was accepted and soon afterwards the Rio Tinto Company was formed. Hugh Matheson was the chairman and financial driving force; his confidence in seeking to acquire the mines had been based largely on advice from Heinrich Doetsch and W Sundheim, general merchants in Huelva with mining interests in the province, David Forbes FRS, a British consulting mining engineer, George Bruce, a British consultant on railway planning and construction, and Dr Romer, an eminent German mineralogist - Forbes and Bruce were appointed consultants to RTC on its formation and played major roles in its development. The success of the British company which took over control of operations at Tharsis from French interests some years earlier would also have been an important source of confidence. The purchase price was the equivalent of £3,850,000. Details of the financial arrangements, a summary of corporate evolution of RTC and some related comments are given in Appendix 7.

The maximum annual production of mined ore at the mines prior to 1873 was 89,700 tonnes under the management of Rua Figueroa in 1863. All previous operations at Rio Tinto had been concerned only with copper production - the new Company was founded to supply the rapidly growing market for pyrites for sulphuric acid production, as well as to continue copper production at the mines, and envisaged shipping 500,000 tpa to northern Europe. From the reopening of the mines by Wolters, operations had been concentrated on what became known as the South Lode and miners were no doubt initially attracted to that deposit by the extensive system of adits driven by the Romans and providing access to the mineralised sections.

The map issued with the new company's prospectus in July 1873, and reproduced in the Introduction as Figure 2, indicated the positions of the lodes of massive pyrites, marked as South, San Pedro, Middle and North Lodes. It was soon appreciated that the plan was somewhat inaccurate - the San Pedro and Middle Lodes did not exist, Spanish engineers having assumed their existence because of the Roman workings on the northern and southern slopes of Salomon; we now know those workings were concerned with extracting jarosite for silver. An important lode to the west of South Lode, the San Dionisio Lode, was not recognised until after operations began.

Forbes, recommended that mining should continue at the South Lode and that opencast working should be introduced; this latter recommendation had previously been made by Anciola and Cassio in 1856. Theodor Blum, the first engineer appointed to the property by RTC as head of the mining and extraction department, was responsible for implementation of this, and other, mining plans in the early years of RTC. Production of the quantities of pyrites envisaged by the company required the opening up of other mining areas and the North Lode was also brought to production by 1880. Detailed examination of the mineralised areas in 1873 no doubt led to some revision of the plans suggested in the company's prospectus - certainly early working revealed a different, but not necessarily displeasing, disposition of the enormous mineral wealth.

In 1873 Rio Tinto had no road or rail links to bring in heavy equipment, nor indeed was much mechanisation in use anywhere in mining. Imports for the mine were taken by barge from Huelva to San Juan del Puerto, thence to a site about 7 kilometres north of Valverde on the Buitron railway and by bullock cart the final 20 or so kilometres to the mines area. The operation depended on mules and donkeys for transport of overburden and ore and, while the mining operation was progressing on that basis, railway connections to the mine were under construction.

Railway transport was crucial to the enterprise. The contract for the railway was undertaken by Clark, Punchard and Company to plans prepared by Bruce. Pier construction at Huelva, again to a design by Bruce, was also entrusted initially to Clark, Punchard but they found the task beyond their experience and the contract was transferred to John Dixon and Company. The design of the system envisaged loading mineral into railway trucks at the mine and hauling the same trucks to discharge to ship loading hoppers on the pier and this required extensive tunnelling to access the mine workings. For almost a century, rail transport was virtually the only means of transporting ore and waste from the mining operations and the levels and layout of the haulage tunnels were of vital concern to all concerned in planning and operations.

In working the South Lode prior to 1873, three Roman adits were used in extraction operations and a fourth for drainage. Since these adits are frequently referred to in reporting on activities at the mines they are listed below:

San Pedro adit, at a level 362.4 metres above Huelva datum and 44.7 metres above the San Luis adit

San Roque adit, 16.6 metres above the San Luis adit.

San Luis adit, the Roman 8th floor, 317 metres above Huelva and 10.3 metres above the railway at Rio Tinto Station.

Cuatro Molinos adit, 31 metres below San Luis adit and used by the Romans for draining the mine; about 2 kilometres long.

The railway entrance to the mine workings of South Lode was driven on the 319.8 metre level and proved more difficult than expected, the rock being porphyry and not slate as had been assumed. Attempts to speed progress by driving also from within the lode encountered difficulties with acid water, delaying work until a Cornish engine with bronze pump fittings was delivered. The tunnel was not completed until November 1876; initially it was known as the Main Tunnel; after 1900 it became the 11th Floor Tunnel.

Delays in completing the Main Tunnel rendered significant exports of pyrites impossible. Forbes had the railway extended from Rio Tinto Station towards the San Roque adit and an inclined plane was built to bring pyrites from the opencast to railway waggons at its foot. This was in operation by the time the railway link to Huelva was completed in July 1875.

Customers for pyrites were concerned about the value of the roasted cinders, which they usually sold to associated companies for the recovery of copper and iron by the Henderson process - Checkland(28) gives an interesting account of the structure of the pyrites burning acid producers of the time. This lead to a need for selective mining to produce the high copper content pyrites required and this proved difficult in the South Lode opencast.

The topmost portion of the orebody, although not as badly affected by oxidation as Forbes had feared, was still relatively low in copper as the Romans and Spaniards had removed the richer parts. Because copper content increased with depth, underground mining was started in 1878 using an improved version of the pillar and stall method used prior to

Figure 11 South Lode Opencast circa 1890

1873. The interval between floors was extended from 10 to 12.5 metres, with galleries, 3m high by 3m wide, and pillars 6m square, as previously but with more attention to accurate vertical alignment.

Underground mining in a body worked from Roman times involved the problems with acid waters already encountered in driving the railway tunnel and the engineers developed the use of bronze fittings and wooden pipe linings to deal with the problems, much as the Romans had done with their bronze shafted, wooden water wheels - extensive use was now made of Cornish engines of the Trevithick design(29).

After completion of the Main Tunnel, overburden and mineral were loaded directly to rail waggons by hand. From underground workings, they were delivered to the 9th floor and fed to waggons in the Tunnel. The underground mine was served by two shafts - the Main or Maestro shaft at the east end for hoisting ore, and the Victoria at the west end, a service shaft.

By 1882, North Lode was in production but production from it and South Lode was inadequate to meet rising needs and work to extend underground operations to the San Dionisio orebody began with extending the Main Tunnel westwards. It reached the western extremity of the orebody by the end of 1885 and for most of its length was in pyrite. The mining method used in San Dionisio was initially the same as used in South Lode. San Dionisio proved to be the most important of the lodes in RTC's operations although not identified by the company until after it began operations at the mines - because of its importance its mining is dealt with in detail in a separate chapter.

By 1893, umderground working at South Lode had reached its limit westward and some 5 million tonnes of pyrite of 2.86% average copper content had been extracted by mining down to the 22nd floor, that is 185 m above Huelva datum. Some 9.5 million tonnes averaging 3.1% Cu had by this time been mined from the opencast and it was decided to extend the opencast westwards to its limit, a decision which implied the destruction in due course of Rio Tinto pueblo.

Julian(30), writing in 1939, refers to the low extraction obtained by the underground mining method used at the turn of the century and the inherent instability of the workings where any pillar robbing had been conducted. In 1856 Anciola and Cassio had suggested use of overburden from their proposed opencast to fill underground cavities; back filling was introduced at the Tharsis mine in 1895 but, despite earlier suggestions from both engineers and managers, it was not introduced until 1900 at Rio Tinto. By that time the underground workings were honeycombed down to the 22nd floor and, with some 20 million tonnes remaining in place in South Lode, the difficulties of maintaining production at the then current levels were sufficiently serious for the Board, in October 1891, to authorise the introduction of the more costly cut and fill methods.

The new mining method was implemented rapidly and effectively, but much of the workings was still unfilled when, in 1907, a torrential rainstorm washed much of the slate footwall of the opencast on to the roof of the underground mine, crushing the unfilled workings down to the 22nd floor, the bottom of the mine. A fire resulted, with sulphur dioxide passing to otherwise unaffected sections of the mine and it was many years before full recovery from the damage was achieved, meantime production fell to around 140,000 tonnes pa.

Average annual mine production is given by Avery(1) as 763,000 tonnes for the period 1876-80 and 1,153,000 tonnes for the period 1881-85, a steady but not spectacular build-up, with exports for the second period, at an average of some 308,000 tonnes pa still substantially below the level of 500,000 tonnes forecast in the prospectus of 1873 and, not surprisingly shareholders were anxious about their investments, as related in Appendix 7. More detailed information on production is contained in Appendix 10.

Figure 12 Heap Roasting circa 1890

CHAPTER 8 Early metallurgical operations of RTC

When RTC took over operations, the Prieto process had been in use for copper production for over 35 years with considerable success, against a decline in the roasting and smelting of ores. Among the difficulties for the smelting route, which have been referred to earlier, were the cost of charcoal and the lack of ore of copper grade sufficiently high to work economically by the smelting route. The Prieto roasting and leaching process achieved higher extraction with lower grade ore and the conditions were therefore greatly in its favour.

Thus continued use of the Prieto process was attractive to RTC for treating the large tonnage of pyrites of lower copper content which was not readily saleable abroad. Although during the Remisa era adverse effects of sulphurous gases on the countryside and its inhabitants had become apparent, protests had seemingly not been so great as to cause concern; indeed two beneficial effects were observed - the inhabitants of Rio Tinto remained healthy during a number of cholera epidemics in the region and the acidity of stagnant waters was such as to inhibit the breeding of mosquitoes and the spread of malaria.

In 1872 the size of the teleras was between 180 and 230 tonnes and much larger heaps were desirable for handling the vastly increased tonnages of ore thereafter. Trials with very large teleras showed that one of 1,500 tonnes, while economising on fuel, took 12 months to roast and some central parts still remained inadequately roasted. A decision was made to limit the size to 500 to 1,000 tonnes.

The calcination ground was located on an impervious slate formation to the east of the railway, as shown in the reproduced 1878 plan (inside back cover). Teleras were built 100 ft long, with a maximum width of 16 ft, between 10 and 15 ft high, and with hemispherical ends. In building a telera, rough flues were made on the levelled ground, using loose stones to give a cross section of about 15 in wide and 18 in high. These flues ran across the telera, at about 15 ft intervals, and connected to two chimneys, one at each end; the chimneys were raised in rough stones in the course of building the telera. Lump pyrites was laid on the floor between the flues and the construction of the telera was then carried out by women and boys who carried the mineral in baskets on their heads, walking up wooden ramps.

On completion, the surface was covered with fine pyrites and allowed to settle for a few days before it was fired by brushwood and logs which had been placed in the flues. Fuel use was about 2.5% of the weight of pyrites. Steam, tar and free sulphur produced during the initial firing condensed on the outer surface, forming a hard crust - where it cracked, it was sealed with fine pyrites. Some control of roasting was possible by damping the flue inlets and chimney outlets, the off gases commonly ran 3 - 4% sulphur dioxide and roasting took about 6 months, depending on the weather. On completion, the telera was broken down and, while still hot, hand loaded into 2 tonne side tipping steel railway waggons. Any unroasted ore and kernels were set aside for later treatment.

The roasted ore was taken to one of three installations of washing tanks - Estacion, Marasmilla and Nerva. The waggons tipped directly into tanks, intially 2 ft deep, 10 to 14 ft wide and 30 ft long, but later deepened to 3 ft and narrowed to 8 ft; each tank held about 40 tonnes of calcines. There followed the complicated washing cycle lasting 7 or 8 days, as described on page 34, after which the washed calcines were reloaded in waggons and dumped on Barbasco heaps, or, as they were originally called, "terreros". Further copper extraction occurred very slowly, as a result of air oxidation and washing, the latter usually only by rain - with ore containing 2.5% Cu it was normal to recover 60% in the tank washing.

A significant discovery made in the early years of operation was that passing the"brown liquor" obtained in the leaching or washing operation through a bed of small sized pyrites before transfer to the cementation tanks reduced the ferric iron content and increased the copper content - reduction in ferric iron content led to a useful reduction in iron consumption. The role of ferric iron in heap leaching was a topic of intense interest, particularly in the role of Heinrich Doetsch in promoting use of patented processes for heap leaching.

The work was labour intensive and very arduous, especially the loading of hot calcines into waggons. About 1,000 tonnes of pyrites was built into teleras daily and, at the end of the cycle, about 800tonnes per day of washed calcines was loaded for dumping. Before 1873 there were 233 teleras roasting a maximum of 65,000 tonnes of pyrites annually; this was increased to 159,000 tonnes in 1876 and again to 300,000 tonnes in 1880 - water supply was the main obstacle to expansion.

This considerable increase in the roasting of pyrites was responsible for smoke damage over an increasing area. On many mornings there was a very heavy blanket of sulphurous smoke over the calcination grounds and in Nerva it was sometimes impossible for people to see their way about. In summer this blanket of smoke, which also carried arsenious oxide, would disperse soon after sunrise but in winter conditions were so bad that work could not start until bells were rung by the Company to summon employees. Occasionally work could not begin before 2 pm - after a serious railway accident, traffic was stopped if the visibility fell to 2 metres!

Apart from the effect on the local people, especially those in Nerva who lived closest to the calcination area, damage to agriculture was considerable and the prevailing wind in the winter months carried the sulphurous smoke towards the best agricultural land near Zalamea la Real. Lack of protection from government and seeming indifference on the part of the Company to the farmers' plight led to the formatiom of an anti-smoke league which encouraged agitators. Finally a meeting was arranged in the Town Hall at Rio Tinto on 4 February 1888. Unhappily, the crowd got out of control and shots fired by the military resulted in serious loss of life - fearing repercussions, relatives furtively buried many of those who died in the "terreros", their skeletons to be found many years later when those heaps were lifted.

This loss of life so shocked Spain that within the space of 24 days the government issued a decree banning open-air roasting. The RTC Board saw this as a direct attack on the Company, despite all the evidence of the damage and distress the roasting operations were causing - it believed the Company had acquired the right to use the process when it purchased the mines and made official protests to the Government. After some resistance, resulting in a dramatic reduction in roasting in 1889, the Government yielded and agreed to continued roasting with a request that the tonnage be reduced.

Whereas Remisa in his operations in the 1830s had shown no concern for the impact of operations on the environment, RTC became concerned and from 1877 experimented with tree planting. Details of the continuing concern about its lands are given in Appendix 8.

CHAPTER 9 Infrastructure for the RTC operations

RAILWAY

The rail system introduced by RTC was essential for export of pyrites, it provided transport to the mines for imported equipment and supplies and, because roads in the province were rudimentary, it provided passenger services for company employees.

From Huelva, the railway, of 3ft 6in gauge and single track except at stations, followed the course of the Tinto river and the distance from Huelva to Rio Tinto station is 83 km. The first 14 km crosses marshy ground, the next 11 is through fertile country to Las Mallas station, after which it begins climbing on its sinuous course beside the river. At the station of Los Frailes, 72 km from Huelva, the track is 152.5 m above sea level; from there the grade increases and at kilometre 79 (Naya station) the track is 257.8 m above sea level. At the Rio Tinto station, kilometre 83, it is 318.6 m above sea level.

The original planning withstood the test of time and few changes were made as the demands of traffic increased. There were five tunnels originally; the Chico tunnel at kilometre 65 was subsequently made into a cutting. The longest tunnel is Salomon, 150 m, and the line leaving this tunnel is supported on a bridge, originally of three spans totalling 67 m. This bridge was washed away in a serious flood on New Year's Day 1888 and was replaced by a single span construction - the repairs cost £25,821!

50 lb rails were used initially and upgrading to 75 lb was gradually carried out. The first locomotives, "C" class, weighed 25 tonnes and hauled 25 hopper waggons, each carrying 10 tonnes, as far as Gades station, 45 km from the mines. From there trains of 50 waggons were hauled to Huelva over the nearly level track, either by two "C" class locos or one bigger loco. In 1905,"K" class locos were purchased from the North British Locomotive Company and these could haul 44 "A" type waggons of 11 tonnes individual capacity directly from the mines to Huelva - in 1920 the trains were increased to 74 waggons.

Two Garrett locos were acquired in 1929 and these each hauled trains of 1,500 tonnes of mineral in 30 tonne waggons. In 1953 six 200 class locos from Robert Stephenson of Darlington were introduced and these pulled trains of 1,000 tonnes. All the locomotives to this time were coal fired steam engines - subsequently diesel locomotives were introduced to service.

The railway system provided passenger transport, notably for the employees living in Zalamea and Nerva, services which operated until 1968.

HUELVA PORT AND PIER

Huelva was a small fishing village until the middle of the 19th century when Ernesto Deligny, who represented French financial interests in establishing mining activities in the Province of Huelva and whose involvement therein has been recounted by Checkland(28), was responsible for studies which resulted in its rapid development for shipment of pyrites. In 1867 the Tharsis company built a pier for shipment of their pyrites on the west side of the estuary at Corrales at the mouth of the Odiel but in 1873 there was no dock or wharfage accommodation for shipping in Huelva and cargo was transported between vessels and shore by barges. Checkland has end-paper maps which show the railways of Tharsis and RTC in relation to others in the region. Bruce, RTC's consultant, considered two schemes - a low level pier requiring lifting arrangements to discharge the contents of individual waggons into the receiving vessels, and a pier with rising gradient allowing automatic discharge of waggons at a height sufficient for the discharged mineral to flow by gravity into the vessels. The second scheme, though significantly more costly, was selected.

There followed considerable debate with the regulating authorities who had doubts about the novel design. Additionally, the bearing characteristics of the estuarial muds meant that considerable attention was needed to achieve satisfactory piling support for the structures. These difficulties were overcome and, after transferring the construction contract, as mentioned on page 40, the pier was completed on 23 March 1876 (and remained in service continuously until April 1974).

It was 1,900 ft long and approached by an embankment 1,500 ft long; 700 ft of the embankment was carried on a steel structure supported by piling because ground conditions would not support a conventional earth embankment. The pier itself was a two-decked strucure, with the lower deck handling inwards goods and export of small tonnage materials - matte, copper, copper precipitate and, in later years, sulphur.

The pier was designed to take ships of 4,000 tonnes but it was able to accept vessels of 8,000 tonnes and even larger ships could have been accepted but the bar at the entrance to the river was the limiting factor. The maximum tonnage shipped in any week was 48,000 and the pier was capable of handling 1 million tonnes pa, double the nominal design figure.

A mineral store, known as the "polvorin", was erected in 1876, with storage for 50,000 tonnes and intended to provide buffer stock to ensure speedy loading of ships. This was especially useful when trains could not reach Huelva, as, for example, when the Salomon bridge was destroyed in 1888.

In 1974 shipment of pyrites was transferred to a new port on the east side of the estuary of the Tinto river. For a time pyrites continued to be hauled by the original railway to Las Mallas, the junction of the old RTC main line with the State system. Since the old main line closed in 1984, all transport of mineral to Huelva has been by road.

HOUSING

In 1860 the population of Rio Tinto pueblo was 1,976 and that of the original Rio Tinto (Nerva since 1891) was about the same. In 1873 the RTC's labour force, including staff, was 950, but this figure excluded those employed by Clark Punchard on railway construction and the South Lode Tunnel. Ten years later, in 1883, the labour force was 9,801. The British company was proving a much more attractive employer than the previous state enterprise and there was a shortage of accommodation to match the inflow; many lived in temporary shelters and nearby caves.

RTC undertook the building of houses soon after it began its operations, firstly at Rio Tinto pueblo and then above it on the Mesa de los Pinos - this new village came to be known as Alto de la Mesa and included a hospital. Some houses were also built opposite the railway station in the mine area, Rio Tinto Estacion.

A small hospital existed in Rio Tinto pueblo at the time of RTC's purchase and it appears to have been under the control of Jesus Alonso Lopez who stayed with RTC until his death in 1894. The first British doctor, J Sutherland Mackay, joined the company in 1879 and moved to Huelva in 1888 to the company's hospital there. He is particularly remembered for his

Figure 13 "The Rio Tinto Mines in Southern Spain"
(reproduced from The Illustrated London News August 7,1875)

interest in football and the founding of the Huelva Spanish Football Club.

In 1880 two further villages were built - Atalaya for the men employed at the San Dionisio lode and Naya for men working at the east end of the property where export mineral was crushed prior to shipment and where the cementation operations were carried out. A further village was built at Dehesa in 1887 for men working at North Lode.

In 1882 houses were built for the general manager and senior staff on a site to the west of the operations with, at that time, a magnificent view to the sierras to the north, hence the name - Bella Vista. In 1891 a club and an Anglican church were built and over the years further houses added. This was the British community and eventually most of the expatriate employees lived there.

With the rapid expansion of South Lode westward, it was obvious that Rio Tinto pueblo would have to be abandoned and El Valle was the town built to replace it. The new church of Santa Barbara was opened there in 1918 and the mines Administration building (Direccion) was finally relocated there in 1938; thereafter Rio Tinto pueblo was virtually a ghost town.

No large Andalusian community would have been complete without its bull ring and RTC built a ring in Rio Tinto pueblo. It apparently attracted undesirable elements in the community and was demolished in 1884. Thereafter the local people were still able to attend bull fights in Zalamea, Nerva and Campofrio.

At Huelva, which had developed as a consequence of other mining operations in the province, there was at first no RTC housing but a hospital was built in 1884. Soon after the turn of the century the company built employee housing, in a development known as "El Barrio Reina Victoria" and another hospital was built here in 1931.

In the 1880s many thought that the port of Huelva could compete with Lisbon and become as large and important; the climate was good and it had obvious possibilities as a seaside resort. A hotel was built, financed by the Matheson Bank, London and Messrs Sundheim and Doetsch, the Huelva merchants who supported Matheson in setting up RTC. It was opened in 1892 with great ceremony coinciding with the 400th anniversary of the departure of Christopher Columbus from nearby Palos. As a grand hotel it was a failure and RTC purchased it in 1894 to accomodate senior Huelva staff and to use as offices; thereafter it was known as Casa Colon.

WATER SUPPLY

At the time of acquisition by RTC, water was obtained from springs, wells and the Huerta de la Cana, all sources in the slate and providing good uncontaminated water but very limited in supply. With the rapid increase in population of Rio Tinto pueblo, the immediate needs for water were met by driving an adit into the slate below the ironstone capping of the Mesa de los Pinos and water from this still flows.

In the operations, wells were sunk at convenient sites to provide water for boilers driving steam operated equipment but the supply from these frequently became contaminated by acid drainage, mainly from heap leaching operations. For the leaching operations, acid water from the Cueva del Lago was used.

Average annual rainfall at the mines, as measured from 1875 to 1974, was 29.1 in, but it varies quite sharply from year to year and seasonally. 1874 and 1875 were years of exceptionally low rainfall and in 1876 two dams were built; both are shown in the 1878 plan. The South Dam was the larger with the smaller Railway Dam below it - this latter dam was filled with overburden from the South Lode opencast in 1909.

Expansion of heap leaching led to the building of a third and larger dam, Marasmilla, completed in 1879 across the Tinto river, and also shown on the 1878 plan.

Marasmilla water was markedly acid and was pumped by steam driven bronze pumps - in 1892 a large Cornish pump was installed to elevate water to a tank, usually referred to at Rio Tinto as a "deposit", Spanish "deposito", from which it was distributed to leaching heaps by gravity. The pump operated until 1920.

In 1882, after a further period of very low rainfall, a decision was made to build a much larger dam at Campofrio, some distance to the north of the mining area. This was completed in the spring of 1883 and had a capacity of 3 million cubic metres and was at an elevation such that it supplied the calcination ground by gravity through a 15 in cast iron pipeline via Quebranto Huesos, at which point a pump was installed to raise water to Rio Tinto pueblo and the operations at higher levels at Cerda.

Until 1900 there was no domestic supply in any of the villages; at that time standpipes were provided. Once again, at this time, shortage of water for operations led to a plan for another dam on the Jaramar river to the east of Rio Tinto and the boundary between the provinces of Huelva and Seville but it proved impossible to obtain the consent of the Seville authorities. RTC reverted to an earlier plan for a dam at Zumajo on company owned land - this had a capacity of almost 2 million cubic metres and was completed in 1907, another year of drought during which the company obtained permission to pump water from the Jaramar river.

Direct water mains to the houses at Bella Vista were provided in 1907 but it was not until RTC had sold out to CEMRT that running water and water borne sewage service were provided to the houses in the Spanish villages. Water borne sewage service was installed in Bella Vista in 1926 after the chairman had suffered a bout of colitis while visiting the mines!

ELECTRICITY SUPPLY

Early operations of RTC were powered by large numbers of individual steam engines and electric power was first used in 1901 at the Bessemer smelter (see page 72) to operate a 60 tonne crane. Electricity was supplied at 220 v dc by a Siemens of Germany 80 hp dynamo, belt driven by a steam engine.

Central generation of electricity commenced on 1 January 1909 at a power station, "Central Electrica", located near the South Dam which provided cooling water, although this site suffered a disadvantage in that coal had to be brought in from Rio Tinto station at an elevation 100 m below it - this required 2 km of railway. Generation was at 3,150 v AC and transmission at 3,000 volts; initially the generators were driven by two Bellis and Morcom triple expansion vertical engines using steam at 190 pounds per square inch and each coupled to 750 kw alternators. Turbo-alternators were subsequently installed; the first in 1911 was of 1,500 kW and the fourth and last of 10,000 kW.

The original boilers were of cylindrical shell type supplied by Babcock and Wilcox, each designed to provide 8,200 kilos per hour at 190 psi. These were replaced in 1928 and 1930 by John Thompson water tube boilers but it was not until 1955 that these boilers were modified to deliver steam at 350 psi for which the 10,000 kW turbo-alternator was designed.

The first connection was for Guillermo shaft at the San Dionisio orebody (see page 53) and in 1912 a 600 hp air compressor was installed on surface between Guillermo shaft and Bella Vista to provide air at 100 psi for the mine - many years later this installation was greatly improved by excavating a large air receiver in the porphyry so that a reserve of compressed air could be built up on night shift when the demand was neglible.

Electric lighting was installed in Bella Vista in 1909 but, for political reasons, public lighting was not used in El Valle until 1920, nor was the use of electricity for domestic cooking permitted in Bella Vista until 1929. Central Electrica was closed down in January 1966, since when electricity has been supplied from the regional electricity utility, Sevillana de Electricidad SA.

ROADS

In 1873 there were no roads for wheeled vehicles entering Rio Tinto, only mule tracks and, with the successful development of the railway, it was not until a very damaging strike in 1920 that any modern road was built. At this time the Government employed strikers to make a road into the mines from the Huelva to Badajoz road, connecting with that road at a point near Zalamea. A second road was similarly built via the Puerto Rubia Pass to Aracena and a third from Nerva to Castillo de los Guardas, to join the road from Seville to Aracena. Roads in the mining area were then developed and by 1930 all operating departments were served by roads.

Prior to this time, the normal mode of transport for staff was walking or by horse, with those living in Bella Vista travelling by train to the main administrative office, Direccion, in Rio Tinto pueblo. The train, known as the "batea" or tray, because the passenger accommodation usually consisted of a platform with seats attached, left Bella Vista at 7.30 am, returned to Bella Vista at noon, returned to the office at 1.30 pm and finally returned to Bella Vista at the end of the working day at 5.00 pm. In later years the batea carried staff further afield, through the central operations area east to the pyrites crushing and classification plant at Naya.

Figure 14 Rio Tinto Pueblo, early 1900s

CHAPTER 10 Mining at San Dionisio and North Lode

SAN DIONISIO

Development of San Dionisio for copper by the Romans seems to have been limited but they undoubtedly extracted silver from the ores below the gossan, in which process some 300,000 tonnes of slag were accumulated nearby. They made four adits into the mass and Palmer(18) noted the connections of these adits to surface shafts, probably for alignment.

Williams(21) emphasised the importance of San Dionisio as follows:

> "the San Dionisio Lode, which is the largest single deposit at Rio Tinto, has a maximum length of 1060 m.; on its upper levels it is continuous with the Eduardo Mass which extends for a further 100 m. to the east. In places the solid ore is unbroken over a width of 280 m., and drilling has proved it to persist for at least 485 m. below the pre-opencast surface. ..... The upper part of the deposit lies between slate and porphyry walls, but where its base has been verified the massive ore is completely enclosed within porphyry walls."

The position of San Dionisio relative to South Lode is shown in Figure 15, a plan prepared in 1939 and hence showing features introduced in later stages of RTC development.
Although at its formation, RTC was apparently unaware of the existence of the San Dionisio Lode, it soon became a major factor in its operations and 5 shafts had been sunk into it by 1882; they were:

Alberto, the most easterly, which did not make contact with the ore because it went through the Eduardo fault which is an unmineralised gap between the South and Eduardo Lodes

Eduardo

Victoria, service shaft for South Lode workings.

San Dionisio, which made connection with the extension of the Main Tunnel in 1882 and was used for hoisting mineral.

Alfonso, another shaft for hoisting mineral.

Another shaft, No 6 Shaft, was being sunk in 1882.

Mining of San Dionisio began using the pillar and stall method as it had evolved in South Lode and by 1900, as described by Julian(30), the mine "had been almost completely chambered to below the 20th floor, and work was proceeding on the succeeding floors down to the 26th, and below that level at intervals of 125 ft. on the 29th and 32nd floors. By this pillar and stall method, carried out at vertical intervals of 41 ft., 20 per cent only of the whole mass of ore could, theoretically, be extracted. .... It was thus urgently necessary that some other method be devised for exploiting the remaining 70 or 80 per cent of the orebodies."

Julian goes on to describe in some detail the efforts made from about 1900 to stabilise the workings by filling from the lower levels upwards in preparation for mining the pillars from the upper levels downward. It was estimated that it would require some 7 million tonnes of waste rock to fill the voids created by the mining to that date of about 18 million tonnes of ore and that would take several years. Julian continues:

> "It was, therefore, decided by the mining engineers in charge to fill with waste the openings in a number of transverse columns extending from the lowest floors to the top floor, and from foot- to hanging-wall, each column so filled being

Plate I.

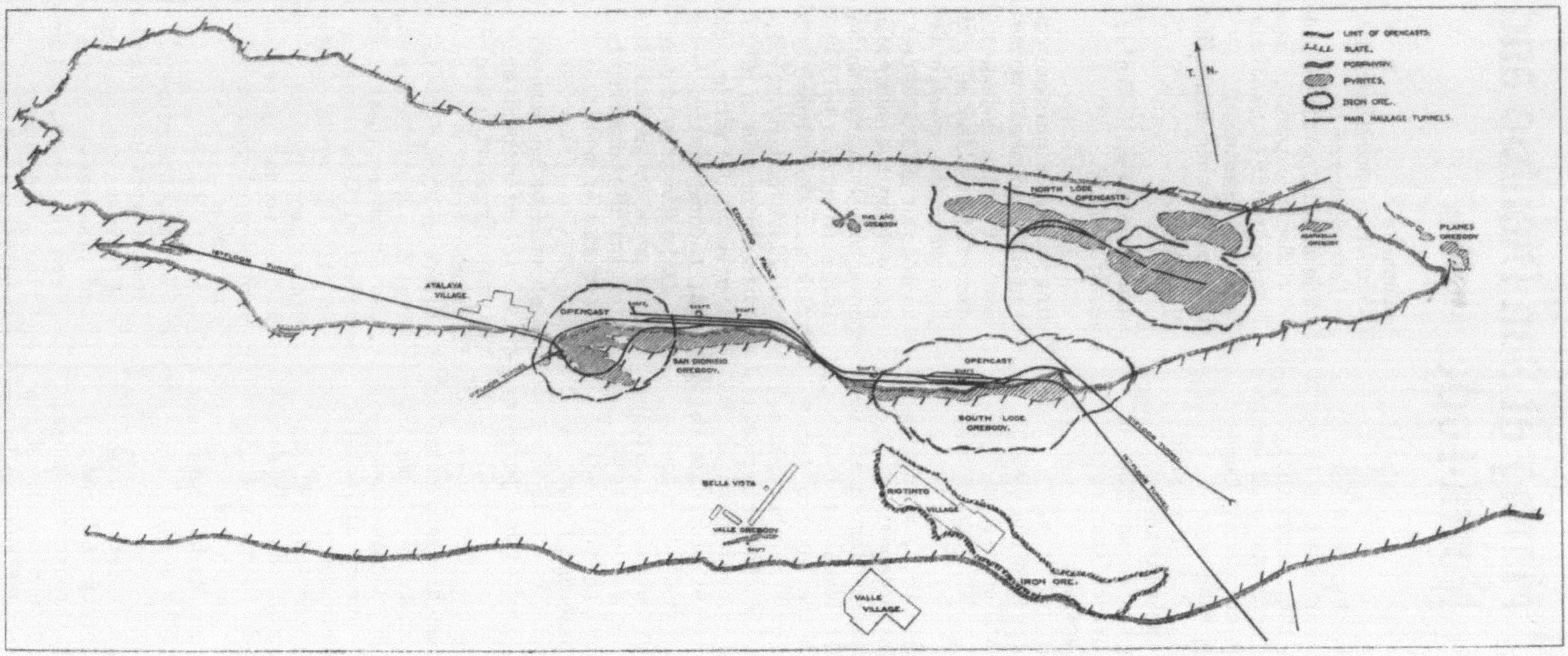

Figure 15 Plan of the Rio Tinto orebodies

about 80 ft. in its dimension parallel with the strike of the orebody. The columns thus filled were at first about 300 to 375 ft. apart. On each floor the longitudinal openings, or gangways, were maintained through the filling by building masonry arches about 7 to 8 ft. high and 5 ft. wide. In each of these columns stopes were started to take out the ore completely from wall to wall and in slices 50 ft. wide, beginning on the second floor from the top and removing first the pillars falling within the stope limits, and then the ore overhead in ascending horizontal cuts about 7 ft. high, and filling with waste rock above.

In practice this did not work out satisfactorily. ....."

The pillars left behind as work extended upwards began to collapse, crushing both the remaining ore, and the masonry built tram roads. Worse, as the ore crushed, it began to oxidise and overheat, creating sulphur dioxide gas in the workings, and raising the already high temperature of the air still further. There was always the risk of fire in such circumstances, paticularly if any timber was left in.

Further modifications in the mining method were necessary and a form of horizontal cut-and-fill was adopted in the 1920s. Julian describes the method in detail - it achieved a very high extraction and less than 8% of the original ore was left in the mined out sections.

Stoping took place on up to three floors, one above the other, to maintain output. The cuts made were filled with waste rock or gossan - placed either by hand or, for a period in the 1950s, by a compressed air "gun". Despite backfilling, in removing three floors, a settlement of up to 9 metres took place, so that access to the working areas was via chutes and crosscuts, in the porphyry rather than the ore, each crosscut connecting to a main gallery running parallel to the orebody.

Working began by taking ore from the weak slate contact side, retreating towards the porphyry gallery at each level. Drilling was by hand machine, and all loading in the 1920s was done by hand, using a hoe-like tool - the rodo - to pull ore into a two handled basket, or espuerta. This basket was then used to carry the ore into 1.5 tonne box-cars running on 2 ft gauge rails. Some attempts were made to use jigger conveyors (especially in conjunction with the "gun") and air-winch scrapers, but these, and Eimco-Finlay loaders were not generally adopted until the 1940s.

Selectivity in mining, to maintain sulphur and copper contents for direct shipping ore, was always a requirement with pyrite production and some of the undoubtedly high cost of the method was justified by the grade control it achieved.

In 1900 the Main Tunnel was diverted around the orebody through the porphyry and renamed the 11th Floor Tunnel. A new shaft was sunk to connect with the realigned tunnel some 250 m north of the San Dionisio shaft - it was named the **Guillermo** shaft, probably in honour of the General Manager of the time, William A Carlyle. The shaft was equipped with electrically driven winding gear, ventilating fans and pumps and did not come into use until 1909 when the Central Electrica power station was commissioned at Rio Tinto.

Two further shafts were sunk about this time. **Alfredo,** close to the Guillermo shaft, extended down to the 32nd floor and was the main dewatering shaft for San Dionisio mine and provided services for men and materials. **Roberto** was a replacement for the Alfonso shaft for hauling ore to the 9th floor for transfer into waggons on the 11th floor.

In 1896, the General Manager, William Rich, considered opencast mining at

San Dionisio but did not press for its introduction and it was not until 1907 that it was seriously proposed to the Deputation from the Board on its visit to the mines. At that time there were three important factors favouring opencast working:

1. Experience with steam shovels introduced to the Dehesa opencast on North Lode in 1904 (see page 57) confirmed that overburden stripping rates could be increased with cost savings.

2. A subsidence near the Alfonso shaft had underlined the difficulties of mining using the underground method then employed.

3. Output from North Lode was falling and it seemed unlikely that underground mining from San Dionisio could be increased sufficiently to compensate.

At first it was decided to work a small opencast in parallel with the underground mine for some few years to assess accurately the relative merits of the two and the Atalaya opencast was initiated by ordering 4 steam driven Bucyrus 3.5 cubic yard shovels from the United States. These were to operate from rail track of the 3 ft 6 in gauge used throughout the RTC operations. These shovels were similar to those in use on cutting the Panama Canal; the story that they were obtained secondhand from that project is apocryphal.

Two shovels were delivered in mid-1907 and one was in operation by the year end. All four were in operation towards the end of the following year and a fifth was ordered. They were used only for loading slate overburden and worked only day shift. Each had a crew of

Figure 16 Steam Shovel Atalaya Opencast

7 men, of whom one arrived early to raise steam by the 8 am general start of work - the team was made up of two drivers, two trackmen, a greaser and a stoker; one driver was responsible for the swivel, the other for crowding the bucket, a division of duties which presumably was understood by those involved!

A team of 7 or 8 was involved in drilling and blasting ahead of the shovel, on a bench of 25 to 30 metres with lifts of about 2.5 metres per blast. Prior to 1912, when compressed air drilling was introduced, all drilling was by hand - 3 men were expected to drill 5 to 8 metres in a 6 hour shift. A shovel advanced about 10 m along the bench each shift, loading about 750 cubic metres of broken slate into 10 tonne waggons, a total of about 170 waggons for the shift. The waste was hauled away by 40 tonne locomotives in 20 waggon trains through tunnels to waste tips to the west.

The north side of the opencast was in porphyry and waste from here was used as fill in the underground mines, which by this time were served by the Alfredo shaft, the older mines having been cut through by the rapidly widening opencast. Blast hole drilling here was closer spaced to give better fragmentation and the loading of the broken waste was done by hand into rail waggons for movement on the 11th and, later, the 16th floors. On the intermediate bench levels, waste was loaded into smaller waggons working on 2ft gauge track for ultimate delivery by the appropriate shafts to the underground workings.

Ore, following drilling and blasting, was moved by scrapers over grizzlies to glory holes on the 20th floor. From there it was delivered by chutes and loading hoppers into 4 tonne waggons on the 23rd floor, hauled to Roberto shaft, raised to the 14th floor, delivered through surge bins to 10 tonne waggons in the 16th floor tunnel and hauled away.

From 1930, steam shovels were gradually replaced by electric shovels mounted on caterpillar tracks. In turn, electric shovels were replaced by diesel machines. The last steam shovel at Rio Tinto, after being used on work outside Atalaya opencast, was finally scrapped in 1961, there being no preservation society to take on what was by then a machine of historical interest.

The first production from Atalaya was 72,000 tonnes averaging 2.82% Cu in 1910; thereafter it became an increasingly important source of low cost production.

Pyrites extracted from San Dionisio and Eduardo Lodes up to 1954 was:

Alfredo underground	26,800,000 tonnes
Atalaya opencast	11,300,000 tonnes
Total	38,100,000 tonnes

NORTH LODE

Preliminary assessment of the orebodies prior to acquisition had concluded that North Lode was the largest of them and when the mining of it began in 1877 it was expected that it would be shown to join with the postulated Middle Lode. At the company's AGM in 1880 the Chairman, Hugh Matheson said: "It is estimated that there is enough pyrites in North Lode to last the Company, at the present rate of extraction, for a hundred years." At that time the rate of extraction was about 900,000 tonnes pa.

It was found in due course that North Lode was not continuous, there was no Middle Lode and North Lode was three isolated and relatively small "masses"; the sub-lodes were:

Mass No 1, later called Lago
Mass No 2, later called Salomon
Mass No 3, first named Balcon del Moro and later Dehesa

Lago lode had been worked by the ancients and the Romans, mainly the former, for the extraction of silver, but not for copper. They drove adits from the north slope of Cerro Salomon, these were expanded to caverns in good ore, and the old workings connected a number of caves, man made or natural, several of which were filled with water. The most famous of the caves were Tabaco and Lago. The Roman workings consisted of shafts and galleries and some of the shafts were filled with slag from silver ore.

Salomon Lode had been extensively worked by the Romans for silver and copper and their main operations probably took place there, a judgement based on the evidence of their working encountered in removal of overburden by RTC. Two major Roman adits , the Upper and Lower Nerva adits, ran in an easterly direction at 55.5 and 24.7 metres respectively above the Rio Tinto station level. It was in Salomon that some of the Roman water wheels were found.

Above the Dehesa Lode there was an area called Balcon del Moro in which many Roman shafts had been sunk through the gossan, either down to the pyrites or to the surface of the porphyry. These shafts or pits were found southward to the Cerro Colorado ridge and to the west as far as Cerro Retamar and the whole area had been extensively worked by the Romans, extracting both silver and copper, the latter from oxidised ores. There were a number of opencasts scattered along the Balcon del Moro, extending from Lago to Retamar, and slag from silver smelting had been dumped in many of them. [After the Cerro Colorado opencast was started by Rio Tinto Patino in the 1960s, galleries and caves were found in the gossan - in section they looked exactly like the model of Roman workings made by the Tharsis company and illustrated in Checkland(28), p 64].

RTC began operations at North Lode by driving an adit from the east along the north side of the Tinto river valley. Initially called the North Lode Heading (its position is shown in the 1868 plan), its level was 34.8 metres above Rio Tinto station and 352.5 metres above Huelva datum. It came to be known as the North Lode or 5th Floor Tunnel.

By July 1880 the adit was north of Cueva del Lago and within 10 metres of the pyrites when considerable ingress of water occurred. Neil Kennedy, the mining engineer in charge of the work, reported on 4 August: "The water level in Cueva del Lago is falling since the heading has reached the lode and there is so much water that the men cannot stay underground the full eight hours." Next week he wrote: "The level of the Lago water has fallen 1.26 metres during this week." In the following week it had fallen a further 2.02 metres and on 25 August he wrote: "The lake is completely empty."

Production of pyrites started that year and the 5th floor tunnel was continued in a large loop into Salomon and Dehesa. In June 1894 underground working of the Lago lode ceased; working had reached the 10th floor, the limit of the sub-lode in depth and the deepest of the North Lode bodies. The operations had produced some 1.3 million tonnes of copper ore averaging 2.04% Cu and about 7 million tonnes of pyrites. Working was by pillar and stall, in which only about 20% of the mineral was extracted. Opencast working of North Lode began in 1901.

Salomon was mined from 1881 to June 1894 and during this period some 900,000 tonnes of pyrites were removed via the 5th floor tunnel. Opencast working started in 1889. In 1904 an inclined plane was installed to remove both pyrites and overburden, the latter being dumped on the old Roman slag heaps and burying the Huerta de la Cana walls.

The Dehesa sub-lode proved to be the shallowest of the three comprising

North Lode and the 5th floor tunnel entered below the ore. Six shafts were sunk into the lode, with underground working starting in 1882 and ending, with Salomon and Lago, in June 1894 - only 550,000 tonnes of ore were extracted.

In 1882 a tunnel was driven from the Main Tunnel, at the 11th floor level and 31.25 metres below the 5th floor level, northwards to the Dehesa lode. This came to be known as the Central Tunnel and was little used until much later. It did, however, have one early use - it connected with an old Roman shaft and waggons were loaded with ancient silver smelting slag, used for railway ballast, from about 1890 to 1931.

This phase of operations at North Lode must have been a severe disappointment to RTC which had initially believed North Lode to be the largest, quite apart from the presumed existence of Middle Lode * being disproved. Some higher grade, quartzitic copper ore was found and proved of some value as a smelter flux.

Dehesa opencast, started in 1900, has already been mentioned as the proving ground for the first serious mechanisation of mining operations and leading to the decision to embark on the Atalaya opencast. The first steam navvies used at Dehesa from 1904 had shovel capacities of 1.5 cubic yards and were designed to load 400 rail waggons, a total of 760 cubic metres, daily and substantially exceeded design performance, hence the ready agreement to embark on the Atalaya project.

OTHER LODES

The Planes Lode is located at the extreme eastern end of the Salomon ridge - at higher levels it is enclosed between slate and porphyry, at lower levels it lies entirely in the porphyry. This lode must have been known to the Spanish engineers although it is not mentioned in their reports and not identified in the RTC prospectus.

When it was rediscovered by RTC is not clear but development work on it started only in 1918 and it was then found to have been worked extensively by the Romans and, relatively recently, by modern pillar and stall methods. There is one reference by earlier Spanish engineers to the finding of an adit of over 1.5 km in length and this adit was lost in the construction of the Marasmilla Dam by RTC around 1879.

Roman operations are assumed to have extracted rich chalcocite and covellite ores found near the slate hanging wall and many remains from their working were found, notably a set of water lifting wheels on the 311 m level referred to in Chapter 2 on page 11. There is convincing evidence that the Romans worked the silver ores at Planes - this evidence is the presence in the area of low copper slags now generally recognised as produced in smelting the jarositic silver ore.

The RTC Board decided in 1921 that "The exploration of the new lode should be pushed ahead at all speed to ascertain the cupreous content of the ore, the tonnage in the lode and its availability for the purpose of annual production." Mineral extraction by underground mining was started in 1922 and by 1933 950,000 tonnes of pyrites and 85,000 tonnes of quartzitic copper ore had been recovered. Some silver ore was also recovered, normally of about 10% Pb and 1,000 to 1,300 gpt Ag and 20 to 33 gpt Au, however one batch assayed 6,800 gpt Ag!

Mining operations ceased at Planes about 1954, by which time a total of 2.2 million tonnes of ore had been mined, and the mine was allowed to flood as the first stage in a plan to recover copper by dissolution in the acid waters. Ore skips were replaced by stainless steel buckets of 5 cubic metres capacity and bailing of the liquor was carried out

* Middle Lode of Figure 2, page 4, soon came to be known as Centre Lode and is thus designated on the 1878 General Plan of Works

in the summer months from 1956 until 1963, by which time the copper content of the waters hoisted was no longer high enough to justify cementation. 1957 was the best year of operation and 825 tonnes of copper were raised in 151,000 cubic metres of water.

An extension of the orebody easterly from Planes was predicted by Dr David Williams and geophysical work began in 1960. Following the identification of anomalies, drilling confirmed the existence of further ore contiguous with the Planes body - this is known as the San Antonio Lode.

Another lode neglected by the Spaniards, although noted by de Cassio in 1867, was the Valle Lode. It was a very small lode, entirely enclosed in slate, marked by patches of gossan and located near the residential area developed by RTC as Bella Vista. The deposit was 230 m long and its width reduced with depth, from about 10 m to 2 m. Underground mining began in 1922 and continued until 1933, with total production of 190,000 tonnes of pyrites. Following the end of mining, copper recovery by cementation of the mine water was practised until 1940 for a production of 525 tonnes of metal.

Three further lodes, all quite small, were found - Mal Ano, Argamasilla and Quebranta Huesos. They are shown on Figure 11. Mal Ano was mined in the early days of RTC and proved a disappointment with only 8,260 tonnes of pyrites with 1.9% Cu being extracted. The other two lodes were found to be similar and of no economic importance.

The total output of pyrites by RTC in its 81 years of operation at the mines (1873-1954) was almost 109 million tonnes and details are given in Table 1 below.

TABLE 1 PYRITES PRODUCTION BY THE RIO TINTO COMPANY LTD

long tons

	Underground	Opencast	Total
South Lode	18,192,713	24,201,493	42,394,206
North Lode	2,754,064	22,928,632	25,682,696
San Dionisio	26,780,834	11,350,589	38,131,423
Valle Lode	193,037		
Planes Lode	2,123,798		
Mal Ano	8,260		2,323,095
			108,531,420

Notes: San Dionisio underground production came from the Alfredo Mine and opencast production from Atalaya.

See also Appendix 12

RTC used long tons; throughout these notes tonnes are used except when reporting detailed RTC production statistics.

CHAPTER 11 Ore classification and preparation for shipment

RTC operations, and those of other pyrites producers in the Andevallo, were unusual in comparison with most mining operations in that a high proportion of the output was sold with no physical beneficiation, although much of it was, in the first half of this century, leached or "washed" to recover much of its copper. Consequently selection and classification of "mineral" was based on both analysis and physical condition. At least 10 categories were recognised in 1888, and recorded by Collins(31), as follows:

1. Pyrites containing less than 0.5% Cu, sold as poor ore.
2. Pyrites mixed with earthy matter and low in copper; it was valueless, hence known as "esteril".
3. Soft powdery or pulverised material of similar composition to poor ore, originally valueless, later called "azufron".
4. Compact ore containing 1 to 2.5% Cu, initially treated on teleras, later heap leached.
5. Compact ore containing up to 3.75% Cu, originally exported.
6. High copper content ore, usually around 4%, occasionally as rich as 12-14%; smelted at Rio Tinto.
7. Quartzitic copper ore, 3-4% Cu. Used as flux in copper smelting; some dumped for heap leaching.
8. Soft decomposed black mineral, high copper, known as "negrillo" and sent to the smelter.
9. Compact mineral containing galena, blende and pyrite; 6-12% Pb+Zn, 3-4% Cu, known as "plomizos" and dumped.
10. Crystals in the form of stalactites or stalagmites with 4-12% Cu and 2-10% Zn; "vitriolas" from which copper was recovered in the washing tanks.

For export ore, sulphur content was critical. Such ore was normally sold on the basis of 48% S, with premiums and penalties for departures from the standard. Typical analyses from this early period were:

	1	2	3	4
Cu	3.06	2.20	4.46	2.25
S	48.98	30.19	47.25	50.00
Fe	41.91	42.86	42.35	41.05
Pb	1.47	-	1.26	Tr
Zn	0.02	-	0.24	Tr
As	1.00	0.92	0.61	0.25
Sb	0.06	0.10	Tr	Tr
Silica	0.28	Tr	2,40	2.60

Ag varied between 17 and 35 gpt
Au " " 0.4 " 0.6 "

Notes:
1 Export ore 1878
2 Ore to calcination 1884
3 San Dionisio 1881
4 North Lode, Lago 1881
All figures, except Ag & Au, in %

Initially pyrites was shipped in lump form and customers did their own crushing, either by hand or mechanically, to meet the needs of their roasting plants. Large sized lumps were required, fines being detrimental to the roasters of the time in obstructing air flow. The shipped ore was roughly minus 8 plus 1 inches, known locally as "tal cual" or so-so.

In early operations, mineral which was less than 1 inch accumulated at site, providing incentive for development of means of realising values from such "smalls".

The demands of the market changed as new roasting equipment, notably multi-hearth furnaces with mechanical rabbling, came into use towards the end of the century. Two crushing and screening plants were used before 1918; one at the Muelle de San Antonio, about 1.5 kilometres from the Rio Tinto station and used later to screen ore for the smelter, the other near Naya station on the railway leaving the mines for Huelva.

In the development of crushing and sizing prior to shipment, it was necessary to reorganise rail transport out of the mine. For many years the 11th floor tunnel had been the main outlet for ore from South Lode and San Dionisio but another tunnel was now needed to deliver to a convenient site for crushing and screening. Several possibilities were considered; the one chosen was at 16th floor level, taking a straight line from Naya village to the east end of South Lode, all in slate, there turning westerly and continuing in porphyry, following the whole length of South Lode to San Dionisio.

Work started in 1912 and the tunnel, 7.6 km long, was completed in 1918. Spoil from the Naya end was used to create the site for a new crushing and screening plant, Zarandas Naya (zaranda is Spanish for sieve or screen). The 16th floor or Naya tunnel then became the main outlet for pyrites and was electrified in 1924 with battery locos and later with trolley locos hauling trains of forty 10 tonne Granby type waggons to the crushing plant.

Delivery into the crushing plant was into four 250 tonne bins. The first unit consisted of a 36 x 18 in Traylor gyratory crusher, and two sets of 42 x 18 in rolls, in closed circuit with screens. This unit could crush and screen 1200 tonnes per 8 hour shift and employed 13 men. The second unit was somewhat larger, having two Traylor crushers, one 5 ft 6 in Symons cone crusher, a 48 in Symons disc crusher and three pairs of rolls - its output was 1500 tonnes per shift employing 17 men.

As designed, Zarandas had 6 product storage bins of 3,000 tonnes capacity and 2 of 2,000 tonnes, discharging directly into main line waggons for haulage to Huelva. There the waggons went directly to the pier for loading to ship, or discharged to a large storage installation of 110,000 tonnes capacity about 1.5 km from the pier. From Zarandas to the pier at Huelva is 80 km and the railway handled a maximum of 4,500 tonnes daily.

Zarandas was designed to supply furnace size, -2.5 in + 1 in, and fines, -0.5 in, but demand for fines increased and, despite modifications, Zarandas was unable to meet the needs and another plant, Lavadora, was brought into service. Much later, about 1970, the Compania Espanola de Minas de Rio Tinto installed a new crushing and screening plant to the east of the original one.

Figure 17 Train Approaching Zarandas Naya from 16th Floor Tunnel 1950s

CHAPTER 12 Hydrometallurgical development

DOETSCH PROCESSES

From the outset of operations, RTC disliked the loss of sulphur values resulting from heap roasting; additionally the problems associated with large scale roasting were recognised. Heinrich Doetsch, of Sundheim and Doetsch, had been appointed a dirctor of RTC in 1873 and was especially concerned about this problem. Also in 1873, M.Joly had patented a process for extracting copper from pyrites using a solution of ferric chloride. By slightly modifying Joly's process, Doetsch obtained a separate patent and the Board agreed in May 1879 to use "his" new wet process - this process eventually became known as the First Doetsch Process; by 1882 Doetsch was receiving royalties from its use.

The process began with crushing the pyrites to about 10 mm and placing the crushed material in heaps 15 m square by 4 m high incorporating a series of dry stone horizontal and vertical flues. Salt, amounting to 2% of the weight of the pyrites, was spread over the flat top of each heap, after which a basin of about 1.7 m square was made on the top, using the ore to form the walls. Chlorine solution was pumped into the basin and trickled through the ore dissolving copper and the solution leaving the bottom of the heaps was collected in launders for recovery of copper by cementation with iron. Theoretically, 1 unit of copper required only 0.44 units of iron for precipitation; in practice it was found that 1.1 to 1.2 units were used, but this was still significantly less than used at that time for precipitating copper from cupric sulphate solution.

Unfortunately the process was not as simple as the above description implies, for many reasons, the chief of which was the difficulty of making the chlorine required for the leaching solution. J H Collins, who was at Rio Tinto from 1880 to 1884, subsequently wrote(39) voicing criticisms of the process and the following extract from his paper illustrates some of the difficulties:

"To produce what was supposed to be a very effective ferric liquor without the use of burnt ore several methods were adopted.

1. Sulphate of iron was heated in a reverberatory furnace, and the "ferric salt" so produced was mixed with the raw mineral before it was lixivated.

2. Sulphate of iron was mixed with salt and similarly heated, the fumes produced being carried into a condensing tower, through which "spent liquor" from the copper precipitating tanks was made to fall, thus producing a "regenerated liquor" which was used for washing raw ore.

3. The sulphate of iron was similarly mixed with peroxide of manganese, and the fumes were similarly employed in producing "regenerated liquor".

"In modifications 2 and 3 the vapour was supposed to contain free chlorine, which, as well as ferric chloride, has a very energetic action on cupreous pyrites. This was, of course, not the case with No 2, though large quantities of hydrochloric acid were present. In No 3, some chlorine was produced at certain stages of heating, but only to a very small extent.

"How little effective this process in its various modifications proved itself at Rio Tinto may be seen by the fact that the great

"Estacion" heap, after three years' washing, still contains more than half its copper in its original insoluble state. Still the copper liberated was obtained very cheaply, and without the serious inconvenience of the sulphur smoke from open-air calcination, which, as already stated, could not be extended however much it was desired..."

By the time those comments appeared attempts to implement the process had been abandoned. Experiments on recovery of the labile sulphur driven off when pyrites is heated were made about this time and were also unsuccessful, the sulphur recovery was only 2 tonnes per day from heating 20-25 tonnes of pyrites and the charcoal consumption was an excessive cost.

Doetsch patented a second process, known as the F C Process, in which a mixture of 3 parts small ore and 1 part calcines was leached after addition of 1% salt and 0.25% ground manganese ore. It seems likely that the idea for this process came largely from a paper presented to the Mining Institution of Cornwall by Brenton Symon, who wrote: "My experiments have conclusively proved that a better result is obtained when burnt ore is used to wash the crude smalls." Symon's connection with Rio Tinto is not known; at that time there was considerable interest in copper dissolution at the several mines in Spain and Portugal exploiting cupreous pyrites.

In 1887 Doetsch received £2,325.15.0 in royalties for the use of this process by RTC. Again, technical staff at the mines were not impressed by the process and in 1887 made experiments to test the significance of adding salt and manganese ore, the basis of Doetsch's patent, and found that straight washing with dilute sulphuric acid gave similar results to the Doetsch F C Process.

The conclusions from this investigation were confirmed by discussion between Johnstone, chief chemist at Rio Tinto, and Mohr, manager of RTC's Cwm Avon smelter in Wales and communicated to the head office. A year previously, in 1886, Sundheim and Doetsch had renewed their contract with RTC to supply 20,000 tonnes of manganese ore for 4 years. Despite the opposition of staff at the mines to the process, it was maintained in operation for as long as Doetsch was a member of the Board - soon after his death in 1894 a Deputation of the Board to the mines reported:

"The Process by which crude mineral is washed with brown liquor obtained from washing burnt ore is used with best effect and has given excellent results not only in increasing copper returns,but in a marked reduction of the consumption of iron. This is the ground work of the Doetsch Process, which was understood to be covered by Patent under which royalty has been paid since 1882. The question of the validity of the Patents is being investigated by the Board.

"The application of a mixture of salt and manganese to the mineral heaps was commenced in 1882 and continued until June 1894. After going fully into the merits of the application with the General Technical Manager and Mines Manager, we are of the opinion that the benefit to be derived from it is not so clear as to render it advisable to resume it and we recommend that the stocks of salt and manganese be sold."

The RTC Board's support of Doetsch in his involvement in the promotion of processes from which he derived substantial benefit, both from royalties and from the sale to RTC by Sundheim and Doetsch of salt and manganese ore, and in the face of dissatisfaction with the performance of the processes by technical staff at the mines, was an episode which can only be judged by the business ethics of the day.

Apart from those commercial issues, insistence on the continued use of Doetsch processes delayed the introduction of "natural cementation" which had been developed at San Domingos when Portugal banned open air roasting in 1878 and introduced at Tharsis in the 1880s. According to Harvey(32), Mason and Barry, managers of San Domingos, offered the process to RTC in 1878, before Doetsch proposed his first process, and while he was a member of the RTC Board. It appears that no serious consideration was given at that time to the process which subsequently became the basis of the company's vast and profitable heap leaching operations.

From 1894, when the use of the Doetsch process was discontinued, the ferric sulphate leaching process was used - this was the process outlined by Brenton Symon. This still required roasted ore to supply the ferric component and roasting of ore for heap leaching continued until 1900 by which time there was a large accumulation of the mixture of calcined ore and raw pyrites smalls, 20 million tonnes according to Avery. These heaps of partly calcined mineral were locally called "morrongos", though the precise origin of the term is not clear. It seems that neither the Doetsch process nor the ferric sulphate process was successful in conserving the sulphur content of the ore in saleable form, which was a prime objective in introducing the former - in later years considerable effort was devoted to recovery of marketable pyrites from the morrongos heaps.

NATURAL CEMENTATION

Methods for accelerating the natural oxidation of pyrites generating an acid liquor in which copper was dissolved had been successfully developed at San Domingos, and Tharsis had started to reduce roasting and introduce natural cementation in the early 1880s. At Rio Tinto experiments were conducted in 1887, by H F Collins, son of J H Collins whose criticism of the Doetsch process is quoted above, and F Johnson, the chief chemist. Because their advocacy of the process was opposed by Doetsch, Collins and Johnson resigned, the latter to join Tharsis.

Despite the pressures from Government to stop roasting in teleras, it was not until after Doetsch's death in 1894 that serious efforts were made to introduce natural cementation - and a contributory factor then was the realisation that copper was being slowly leached from dumps of ore which had been set aside as too low in copper content to justify treatment. A report by the general manager, William Rich, in 1895 gives important details of the plans to introduce the "new" process:

> "The present proposal is not to calcine any mineral averaging 1 per cent copper in the air, but to lay it down in a heap for natural oxidation and is based on the following consideration. During the years 1889 and 1890, in consequence of the forcible restriction of open air calcination, large heaps of poor rough ore, averaging 0.75 per cent copper, were formed and allowed to decompose naturally. For the first two or three years the action seemed to be slow, but since then there is undoubted evidence of active decomposition, and, although from the situation of the heaps, no absolute measurements can be taken of the liquors resulting from the washings, there is strong evidence to show that probably 20 to 25 per cent of the copper contents remaining in the heaps is being recovered each year.
>
> "The proportion of copper recovered by calcining in the open air is usually taken as 70 per cent of the total, and the results obtained in practice agree with this, but that is when dealing with an ore averaging about 2 per cent copper, the washed residues of which

usually contain about 0.6 per cent. No extended results are available of the value of the washed residues from ore containing only 1 per cent, but from the various observations made from time to time, there is good reason to suppose that they probably contain not less than 0.5 per cent, in which case the copper recoverable from the poorer classes of ore would not exceed 50 per cent of the total, which, I am inclined to think, is about the result obtained on a large scale.

"From an examination of the extraction returns for the present year, it is probable that about 200,000 tons of rough ore can be selected during 1896, averaging about 1 per cent copper, and therefore containing a total of 2,000 tons of copper. If this mineral were calcined in the open air, 50 per cent or 1,000 tons of copper might be expected to be recovered at the end of 12 months, whereas if put into a heap and allowed to decompose naturally, about 25 per cent or 500 tons of copper might be expected to be recovered at the end of two or three years. The total difference, therefore, would probably not be more than 500 tons of fine copper, and I am inclined to think, would very likely be less, as I believe the proportions given above are less in favour of the present proposal than the actual results will prove to be. A point not to be lost sight of is this, that whereas after calcination and washing, the residue yields up its remaining 50 per cent of copper extremely slowly, the uncalcined heaps, on the other hand, would continue to decompose for several years at an increasing rate, and consequently the total copper recovered from these heaps would be for several years an increasing quantity."

Rich continued to show that with a reduction of 200,000 tonnes annually in open-air calcination, output of copper could be maintained over a transition period until all open-air calcination could cease. He continued:

"It is perhaps unnecessary to dwell on the excellent effect that will be produced throughout the entire Province and even in Madrid, when it becomes known that the Company has reduced the calcination from 300,000 tons per annum to 100,000 tons, besides which the area damaged by smoke will then be undoubtedly reduced, and consequently the smoke indemnities in the outlying districts will also be reduced.

"A very considerable saving will be effected in the traffic expenses. It is well known that the very roughest usage to which the waggons are subjected in the whole establishment is in carrying the hot calcined mineral to the tanks, and then removing it after washing, dripping with acid liquors, to its ultimate destination. When the new arrangement comes into operation it will result in diminution of 400,000 tons of the worst kind of traffic (200,000 tons to the tanks and 200,000 tons from them), thus effecting a very large saving in repairs and setting free the waggons and locomotives for service elsewhere.

"The quantity of ferric iron ... in the liquors for cementation will ultimately be lessened. The excessive consumption of pig iron during recent years at Naya was due to the large proportion of ferric iron in the liquors, and although this has happily been reduced in consequence of washing the heaps with less water, the consumption of iron there is still 2 tons per unit of copper produced. At Cerda it is about 1.5 ...."

Heap Leaching

Rich's report and recommendations were accepted and the first of the natural leaching heaps or "terreros" was put down in 1896-7 at Correlejos, an area between Cerda and Naya. It contained some 290,000 tonnes, which proved to be far too large, with chimneys similar to those for teleras, but now to provide natural ventilation to oxidise the pyrites, and with drains carrying the leach liquor to cementatation tanks at Naya. Rich's predictions were fully realised and from 1900 no more low copper content ore was roasted in the open air.

The Tharsis Copper and Sulphur Company was well established before RTC was formed and there was naturally considerable rivalry between them. However mutual interests in the orderly marketing of pyrites had reconciled them to a considerable extent and in 1899 a party of RTC directors accompanied by Rich and Carlyle, who succeeded Rich the following year, visited the operations at Tharsis. At that time Tharsis had been operating only natural cementation for 10 years and had improved leaching performance considerably in that time - they had shown that a reduction of copper content from 1-1.5% in the ore to 0.2% in "washed ore" could be achieved in three years.

Key factors in the process which they had demonstrated were crushing the ore to -50 mm, initial washing with acid liquors, controlling heap temperature to about 50 degrees C, by ventilating and intermittent washing, and passing leach liquor through a bed of crushed pyrites to reduce ferric iron content. RTC practice benefitted from the visit, following which there was considerable experimentation at Rio Tinto on heap height - it seems that Tharsis had, perhaps fortuitously, chosen a near optimum of 10 m - the lower heaps tried at Rio Tinto were abandoned and the height used subsequently ranged from 7 to 10 m.

A very detailed account of operation of heap leaching and cementation was published by Taylor and Whelan(33) in 1942. Heaps, after the first of 290,000 tonnes at Correlejos, were normally of about 100,000 tonnes until 1928 when 50,000 tonnes was adopted as standard. They were built initially on a base of large stones but, again from 1928, selected large lumps of pyrites provided the base which was laid out carefully to provide channels for entry and circulation of air. Run-of-mine ore, containing a substantial plus 9 inch fraction, was used, the benefits of crushing and sizing shown in the Tharsis work presumably not being sufficient to offset the added cost and complication entailed; there seem to have been late doubts about that judgement since a trial heap made in 1934 using crushed pyrites leached rapidly and efficiently.

It originally took about 5 years to extract most of the copper but by 1932, with more careful control, especially of temperature, leaching time was reduced to 24 months and it took 3 months to build a heap and 3 months to remove the leached ore to the crushing and screening plant. Minus three-eights of an inch was sold as "washed ore", sometimes the fraction minus one and a half inches, if sufficiently low in copper was crushed and also sold as washed ore; the rest was returned to be incorporated in new heaps for further leaching. About 20% weight loss occurred during the cycle of operations.

Heap leaching was carried out in three areas - Tejonera, Cerda and Naya. The first two areas were used in the earlier years and Naya became the main centre after the 16th floor tunnel came into operation in 1918. An average of about 2,000 tonnes of ore per day was sent to the heaps; the maximum was about 4,000 tonnes. Ore was delivered from the mines by railway waggons and all transport to and from the screening and crushing plant was also by railway waggons, thus the heap leaching operations involved a major railway network with 40 km of track.

Originally, most of the ore came via the 11th floor tunnel which was at a level 319.8 m and above all the terreros; when the 16th floor tunnel came into use at 252 m it was too low to provide direct distribution to the heaps and an inclined plane was built at 2 km from the tunnel mouth - this could raise ore to four levels at the Naya terreros; 265, 278, 292 and 310m. The inclined plane sloped at 35 degrees and two 7 tonne skips hoisted 200 tph, dumping into 300 tonne bunkers, one at each level, for distribution to the heaps. Over 400,000 tonnes were hoisted annually, at an average power consumption of 0.45 kWH per tonne. The 11th floor tunnel continued to serve Cerda terreros, usually with ores of below average grade.

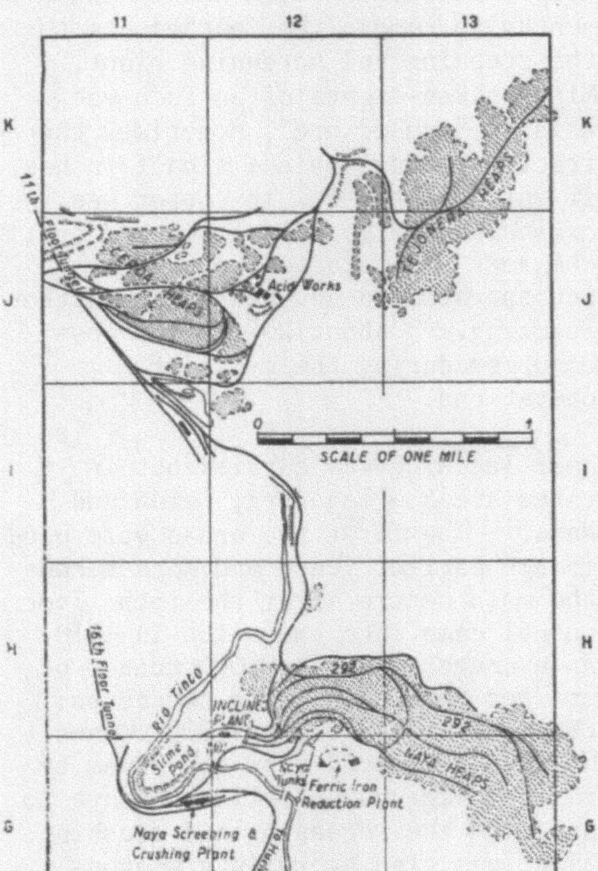

Figure 18

General Plan of Leaching Grounds

Between 1903 and 1953 almost 20 million tonnes of washed ore was sold; after 1951 very little freshly mined mineral was heap leached.

Over many years it had been observed that the rate of formation of ferric sulphate in the heaps was greater than could be accounted for on the basis of normal chemical reaction. In 1961, Dr Paul Trussell of the B C Research Council, Vancouver, visited Rio Tinto. He diagnosed bacterial acceleration in the leach process and subsequently identified thiobacillus ferrooxidans in the liquors at the mines. This was one of the first confirmed identifications of the role of micro-organisms in leaching processes - conditions in the mine waters at Rio Tinto, with pH 2 to 3 and temperature about 35 degrees C are near ideal for thiobacillus oxidation of sulphide and the process has been operating since the sulphide bodies were exposed to air and moisture; certainly since Roman times.

Water Supply

As mentioned in Chapter 9 on infrastructure, supplying water for the leaching heaps was a major consideration; the natural leaching heaps were just as demanding of water as were the operations of Prieto's artificial cementation process. Wherever possible acidic mine waters were used but the fresh water requirement in a typical year, 1932, was about 4 million tonnes, additional to some 1.7 million tonnes pumped to the heaps from mine workings.

Circulation of liquors by pumping involved special materials of construction. Reference has already been made to the introduction of phosphor bronze for pumps in the mines. When centrifugal pumps were introduced to the heap leaching operations they used bodies of antimonial lead with phosphor bronze impellers and steel shafts protected by bronze sleeves, then in 1930 impellers and shafts of stainless steel were introduced. The main distribution pipes were of cast iron, lead lined, 16 to 18 in in diameter, with secondary distribution through pipes of similar construction 6 or 8 in in diameter. Moving pipes was an extremely arduous job since no mechanical lifting gear was available for most of the work.

The four largest circulation pumps had the following characteristics:

Size	8 in
Stages	4
Capacity	200 cubic metres/hr
Head	50-60 m
Motor HP	60-85
RPM	1450
Voltage	3000

Control of rain-water was a problem. Occasional heavy rainstorms occur at Rio Tinto and an inch of rain was roughly equivalent, when the heap leaching was in full operation, to 100,000 cubic metres or tonnes of water. This amount could normally be comfortably absorbed in the heaps over a day or so but greater falls resulted in diversions to the river. Maintenance of storm drains was of great importance so that diversions could be made without serious interference.

Cementation (Precipitation)

Precipitation was carried out at the three heap leaching sites at the turn of the century, much as it had been at Planes and Cerda at the time RTC acquired the mines. Liquors from Prieto's process had been treated at Planes, the Doetsch process liquors were treated at Cerda.

Once the natural cementation process was established, precipitation was confined to Cerda and Naya. The installation at Naya was substantially rebuilt in 1925; after 1927 only the Naya tanks were used and liquors from Tejoneras and Cerda were transferred there by wooden launder at the 292 m level and recirculated on the heaps there; their copper content was no longer separately recovered.

Liquor leaving the heaps was collected in several dams, from which liquor was drawn off through a series of thick walled lead pipes operated as syphons with wooden plugs inserted or withdrawn to stop or start flow. The liquors contained about 2,000 ppm of ferric iron. In early operations this was reduced by passage through lump pyrites, as practised in Figueroa's time and reconfirmed by Tharsis work in the 1880s.

In 1929 reduction by sulphur dioxide was introduced; the reaction, which produces ferrous sulphate and sulphuric acid, is very slow in cold acid solution. Sulphur dioxide was produced by roasting "furnace size" pyrites, about 2.25 in, in lump burners; 40 of these were installed in two sets of 20 each. The burner flues connected to an absorption tower, 6.4 m high by 2.2 m square, lined with lead sheet and filled with heat and acid resistant packing. A lead covered and lined fan sucked the gases upward against liquor descending in the tower and discharged the gases through a stack. Liquor was fed to the tower by gravity from No 2 dam and the sulphur dioxide charged liquor returned through lead pipes below the surface of the liquor in No 3 dam.

In 1937, to hasten and complete reduction, a submerged wall containing 8,000 tonnes of pyrites was built about midway across No 3 dam, which became dams Nos 3 and 4.

CHAPTER 13 Pyrometallurgical development

By 1872 the smelting of cupreous pyrites in blast furnaces had virtually ceased because of the shortage of charcoal. The size of the blast furnaces was limited to the water power available to drive the bellows; very often the bellows had to be worked manually. The advent of the RTC using steam power revived interest in smelting pyrites.

Alexander Hill was responsible for starting the first blast furnace for RTC in 1879. It was sited to the west of the inclined plane near the San Luis adit and in 1879 it produced 39 tonnes of matte of 35% Cu.

Feed to this furnace was described by Collins(31) as "The rich cupreous pyrites and quartz mineral, selected from the Calcination Heaps, ... roasted and smelted together with selected kernels from the Teleras and washing tanks and certain inferior copper precipitate washings".

In December 1880, the Board recommended that further blast furnaces should be erected using, instead of slate, the best available Stourbridge firebricks for furnace lining. By the end of 1884 eight furnaces had been built and the smelter was known as Fundicion Mina.

A second smelter also came into operation in 1884. Located below the South Dam wall in the area known as Huerta Romano, it was known as Fundicion Huerta Romano and consisted of a further twelve furnaces.

Matte from smelting was usually shipped to South Wales, sometimes to Liverpool; both places operated metal markets where smelter-refiners purchased offerings. The unfavourable terms on which its matte was sold led to RTC leasing its own smelters; first in 1884 at Cwm Avon near Port Talbot and second, in 1887, the Grange Metal Works at Jarrow.

Figure 19 Cerda Precipitation Tanks 1900

At the mines, the cupreous pyrites and "quartz mineral" (quartzitic copper ore as defined on p 59) were roasted together before smelting, the aim being to remove sufficient sulphur to produce a matte containing between 31 and 38 per cent copper. This roasting was done in the Cerda area in what was called No 2 Teleras. Again according to Collins:

> "The richer and most siliceous minerals ... were made into small teleras of 200 to 300 tons and roasted in the open. The method of construction is somewhat different to that employed for ordinary roasting and more fuel is required. The ground chosen is covered to a depth of two feet with brushwood and several bundles are placed in a sheaf in the centre to form a chimney when burnt out. Ore is then piled around and the wood kindled, no layer of fine mineral being placed on the outside, as in the case of ordinary heap calcination. Such a telera will continue to burn for about three months."

Roasting temperature was higher than for artificial cementation, copper sulphides were oxidised to cupreous oxide and any copper sulphate formed as an intermediate was decomposed to the oxide.

In December 1882, William Edmonds came from a South Wales copper smelter to take charge of smelting and refining and was responsible for starting up Fundicion Huerta Romano in 1884-5. He suggested that the ore should be roasted in teleras of the same shape and size as those used for artificial cementation, believing that roasting time would be reduced and, with increased sulphur elimination, higher matte grade would be achievable. Larger roasting heaps were apparently used, possibly with improved fuel consumption, but matte grade was not improved. In 1886 he returned to South Wales to supervise smelting and refining at the RTC's recently acquired Cwm Avon smelter - shortly after taking up that appointment he was complaining of the quality of the matte received from the Mines!

Fundicion Mina 1879-1899

No drawings of the small blast furnaces have been found but they were probably similar in general design to those used earlier at the Swedish Falum Smelter in 1833 and described by Bernaldez(19). A sketch, probably made in 1884, shows eight furnaces built in pairs. According to Gonzalo y Tarin(7), the smelter had two piston blowers, each of 120 hp, and steam was provided by two Cornish boilers, one of which was a stand-by; it is believed the blowers were driven by Trevithick vertical beam engines.

Each furnace was capable of smelting 12 tonnes of burden per day. Slag was tapped into small conical cast iron pots carried on two wheels and was dumped in front of the smelter building. Matte was cast into sand beds and then broken up for shipment to South Wales.

Between 1880 and 1889 matte containing some 15,000 tonnes of copper was produced from this smelter - production and burden details are summarised in Appendix 9.

Fundicion Huerta Romano 1884-1901

Originally planned, in 1884, for ten furnaces, twelve were built and operational by the end of 1885. They were similar in design to the furnaces at Fundicion Mina, with blast air provided by one large blower driven by a 220 hp steam engine. Coke requirements for smelting were now significant, amounting to 2,000 tonnes per month.

Following a successful trial with a circular water-jacketed furnace at the Company's Cwm Avon smelter, two such furnaces were installed early in 1888, followed by another ten in 1890. No details of the furnaces

Figure 20   Bessemer Smelter

have been found in the RTC records but they were probably of 3 ft diameter at the tuyeres, a size of furnace in common use around this time. Air was supplied to each furnace by its own Roots blower driven by a 13 hp steam engine.

In 1888 the total copper output from Rio Tinto was 5,792 tonnes in the form of matte and 12,730 tonnes in precipitate. Both smelters were working and the numbers employed in the metallurgical departments were:

	Men	Women	Boys
Teleras	420	11	34
Washing calcines, leaching, etc	280	46	94
Doetsch&FC Process at Planes	110	2	11
at Naya	242	5	14
Roasting ores for smelters	95	23	50
In the smelters	352	10	43
Totals	1,499	97	246

Smelting Precipitate

RTC scrapped the spleiss furnace which had been used since installed by Remisa in 1831 and replaced it by the more common type of reverberatory furnace with an open hearth. As early as 1879 it was suggested that the precipitate should be melted and cast into ingots of black copper for export to the Swansea and Liverpool markets. However it was not until March 1883 that William Edmonds was authorised to engage two Welsh refiners. Initial instructions to send the ingots to Garston, near Liverpool were changed in favour of shipment to Wales, probably in anticipation of taking up the lease on the Cwm Avon smelter.

Bessemer Smelter 1901 - 1914

Before 1900, the time to make refined copper from the Rio Tinto ores must have been between 4 and 5 months. The ore was roasted in No 2 Teleras, an operation which took 3 months, then smelted to matte and the matte shipped to South Wales. The years from 1880 were ones of rapid progress in copper smelting, most notably in the successful application of the Bessemer converting principle to treatment of copper matte.

The rapid development of this process, following the initial demonstration by Manhes, must have impressed the RTC Board and a small converter was installed at Cwm Avon. Tests there showed that matte could be converted directly to copper and the method was much quicker and cheaper than the classical Welsh process. In addition, arsenical copper precipitate could be added to the converter with elimination of most of the arsenic.

In 1899 RTC placed an order for a new smelter, at an estimated cost of £60,000. Construction began on a site along the Tinto river, just below the Marasmilla dam, and was completed in November 1901. At that time Fundicion Huerta Romana ceased operation.

The new smelter was housed in Victorian-style stone buildings (which subsequently became the Mines Workshops) and the main equipment was:

1. Two water jacketed rectangular blast furnaces (a third was added later) each 12ft long, 3ft 7in across the tuyeres, 13ft 4in high and 4ft 4in wide at the top. Each side had 3 vertical water jackets, each with 3x4in dia tuyeres. Above these jackets were 2 horizontal ones. One of the end jackets was fitted with a water cooled spout for continuous discharge of slag and matte into a circular settler/forehearth of 35-40 tonnes capacity.

2. Six converter stands and twelve horizontal converter shells, each 8ft long and 5ft dia; for ease of relining made in two sections, bolted together. These were fitted to or removed from the stands by a 60 tonne, electrically driven, overhead crane; they were turned on their stands by vertical rack and pinion gear, operated hydraulically.

Flue gases from both blast furnaces and converters passed into a settling chamber, thence to a mountain flue, much of which still remains.

The design for the converters came from the Copper Queen smelter, Bisbee, Arizona and they were known locally as Bisbees. No doubt it was James Douglas who advised RTC and arranged for two men from Copper Queen to start up the smelter.

More detailed information about the Bessemer smelter and its operations is contained in Appendix 10.

Figure 21 "Bisbee" Converter

Pyritic Smelting

In December 1900, Charles Fielding, an RTC director then in Huelva, wrote a memorandum about a new method of smelting cupreous pyrites in blast furnaces using little or no coke and utilising the heat of reaction from the oxidation of sulphide. He realised that it was far too late to make alterations to the Bessemer smelter, then under construction, and suggested that tests be made once the Bessemer smelter was operational.

In 1904, a third blast furnace was erected at the Bessemer smelter for the tests and these must have been encouraging since, in December of that year, permission was given to design and erect a new smelter. This was blown in in April 1907 - originally called the Bessemer Extension, it later became known as Fundicion Pirita. With many modifications, it operated until 1970.

The plan called for the erection of six water-cooled blast furnaces to smelt a charge of high sulphur content cupreous pyrites and fines-free silica. This burden was designed to promote an even distribution of the air flowing up through it and the charge column was to be about 12 ft, a height assumed to give good heat exchange between the hot gases produced at the focus of the furnace and the cold charge descending, thus minimising coke consumption. The low copper content matte produced was to be sent to the Bessemer Smelter to be upgraded by resmelting before converting.

However, it was not until 1917 that pyritic smelting according to this plan was practised. Until that time the blast furnaces were used for scavenger smelting of small size cupreous pyrites, containing a large percentage of -6mm fines, of which a large tonnage had accumulated because they were not saleable in the export trade. A means of realising value from such fines had been sought for some years.

The site chosen was an area of flat ground, about 2 km from the Bessemer Smelter, at the foot of the Sierra Madronal and again convenient for the construction of a mountain flue. This flue was somewhat larger than that for the Bessemer Smelter and had an internal cross section 4.5 m square and a total length of 380 m, rising to a height of 159.3 m above the furnace feed floor. In practice, the flue gases were diluted by adventitous air to around 0.3% sulphur dioxide at discharge and, by the standards of the times, caused little pollution.

The six blast furnaces were housed in a structural steel building, with space left for a converter aisle. A boiler house, built in stone and with six Lancashire boilers, provided steam to drive blowers providing blast air. Initially there were four Parsons turbo-blowers, each capable of 660 cubic metres free air per minute at 20 mm Hg pressure, and three other blowers, each 450 cubic metres at 103 mm Hg. In 1917 a German AEG turbine driven blower was installed and in 1929 the first electrically driven blowers were installed - four Oerlikon turbo blowers, each driven by a 450 HP motor and capable of 480 cubic metres at 735 mm Hg.

Electric power was not used at the smelter until 1914 and then it was only used for two cranes transferred from the Bessemer Smelter when that shut down.

In operations smelting cupreous fines, the pyrites contained between 3 and 4 % Cu, and the siliceous copper ore used as flux, contained a similar copper content and 14 to 15 % S. With this small sized burden, the smelting height above tuyeres was only about 5 ft, compared with the designed 12 ft for pyritic smelting, and blast pressure was 83 to 100 mm Hg. Each furnace smelted about 300 tons of burden per day and the average number of furnaces in blast was just under four.

Appendix 11 lists the charge smelted in 1913 and gives some details of operating practice at that time. Further details of smelting practice are given by Potts (34).

In 1913 a structural steel building was erected immediately in front of the blast furnaces and the converters were installed by 1914; these were two 12 ft Great Falls type and two 14 ft long by 8 ft 5 in horizontal converters. After experience with both types, the Great Falls was found to perform better on the low grade mattes and a further three were installed.

Air for the converters was supplied by two steam driven blowing engines supplied by Fullerton, Hodgart and Barclay of Paisley at a cost of £5,250 cif Huelva. They were impressive machines - each had two steam cylinders, 17 and 34 in dia, in tandem, driving two 36 in dia air cylinders at 100 stokes per min, with stroke length 36in. Superheated steam pressure was 10 kilos per square centimetre and each engine supplied 230 cubic metres of free air per min at about 2 kilos per square centimetre. Converting operations have been described by Potts (35,36).

True pyritic smelting commenced in 1917. It was thought that high blowing rates would be needed and the installation of the AEG turbine driven blower in that year was the prerequisite - it was capable of 2270 cubic metres free air at 100 to 130 mm Hg pressure. This blower had been ordered in 1913, it seems to have arrived at Huelva around the start of World War I and clearance through Customs was delayed by the German consul there.

Efficient pyritic smelting required careful screening of the pyrites burden at one quarter of an inch. Good quality silica and limestone of about 4in and free of fines were also needed and were delivered by rail from the Muelle San Dionisio screening plant near the Bessemer smelter.

Coke was reduced to 0.5 % of burden and, for one short period, eliminated. Campaign life for the blast furnaces under these conditions was short - 5 to 7 days. Increasing coke to 1.8 % of burden increased campaign life to 29 days - these latter conditions were adopted for subsequent operations.

In this time of hostilities between Britain and Germany, there were efforts on behalf of German interests to hinder the development. The smelter superintendent, Arthur J Caddick, wrote later:

> " Throughout the experiments, owing to enemy propaganda, difficulty was experienced with labour, so much so, that alterations in practice were published in the local Sindicalist pamphlets, comparing the old amount of fuel used with that which we were using. At times the campaign reached such an intensity that over some months I was frequently threatened on account of what was termed "El conflicto de carbon". Granted the work was more arduous and the furnaces had to be cleaned out more frequently..."

Successful pyritic smelting required good quality pyrites and quartz - typical analyses were:

	Pyrites %	Quartz %
Copper	4.44	1.40
Iron	40.18	12.54
Silica	2.15	65.20
Sulphur	47.37	14.16
Arsenic	0.70	0.38
Lead	1.88	0.79
Zinc	1.75	0.80

Difficulty in supplying such feed led to the cessation of pyritic smelting in 1924, following a meeting at the smelter between Walter Browning, general manager, R E Palmer, chief mining engineer, John Black, chief accountant, and Herbert Potts, smelter superintendent. Palmer's comments to Potts are said to have been: "The smelter will have to take what is sent to it and like it, because we (the mining engineers) haven't time to be running around looking for prize packets of fancy ore tied up with ribbons". Obviously cupreous pyrites of the quality sought by the smelter had become increasingly difficult to extract.

Operating results during the pyritic smelting era are given below; column 1 gives overall average figures but excludes two periods - 1921, when pyritic smelting was not practised because of the need to smelt considerable tonnages of precipitate and for which figures are given in column 2, and a 6 months period in 1922 when no precipitate was smelted and coke usage was thereby minimised - figures for this period are give in column 3.

	1	2	3
Average number of furnaces in blast	3.55	2.20	2.65
Tonnes of pyrites smelted			80,025
Copper content %			2.92
Tons of quartz			29,996
Copper content			3.02
Total tonnes of burden/furnace/day	322	317	361
Total burden, tonnes			132,970
Coke, % on burden	1.83	4.76	0.9
Precipitate smelted, tonnes per day	11.5	72.3	nil
Matte produced, tonnes per day			83
Matte Cu %	22.1	32.0	20.4
Slag Cu %	0.55	0.64	0.63
Blister produced, tonnes per day	27.3	46.0	25.5

The average length of a furnace campaign was 29 days, the height of charge in the furnace was 12 ft, each furnace used 400 cubic metres of free air per minute with blast pressure 110 mm of mercury.

After January 1924, the smelter practised semi-pyritic smelting; the sulphur content of the cupreous pyrites was lower and more coke was needed. In the period 1924 - 29 operating results were as follows:

Average number of furnaces in blast	2.27
Average daily burden per furnace, tonnes	342
Coke on burden	3.45%
No 3 precipitate smelted per day, tonnes	4.57
Copper in matte	20.08%
Copper in slag	0.41%
Average production of blister, tonnes per day	40.1

Comparing pyritic and semi-pyritic smelting on the basis of burden smelted per square foot of hearth area per day gives figures of 4.47 and 5.44. At this time the blast furnaces were 18 ft long by 4 ft wide - each furnace had 6 steel jackets per side, 2 back jackets and a bronze breast jacket fitted with a steel spout; all the jackets were watercooled. The height, from sole plate in the hearth to feed floor, was 17 ft 4 in, settlers were 14 ft diameter by 4 ft 10 in high.

Smelting Flotation Concentrates

The presence of copper in the country rock at the contact with the pyrites was first seriously examined by H H Knox with exploratory drillings in 1910 but it was not until 1926 that the applicability of flotation was tested in a pilot plant at the Lavadora site and the decision to build a flotation concentrator to treat the chloritic ore was taken.

Sintering was an essential first stage to smelting the fine sulphide concentrates and two standard Dwight-Lloyd machines were installed to the west of the mountain flues and some 450 m east of the new concentrator. Concentrate pulp, at 72 - 74% solids, 80% -325mesh, was pumped through a 4 in steel pipeline to the sintering plant where it was filtered on a Dorrco drum filter to 8 - 11% moisture.

The flotation concentrator began

operations in July 1930 and, in a period of very depressed copper prices, was shut down in March 1932 - flotation concentrates were not produced again until 1942. The sintering plant operated only until June 1931 and failed to achieve the expected performance. Concentrates containing 17% Cu and 42% S needed dilution with inerts to reduce sulphur to around 18% to avoid overheating - the mixture used was:

Concentrates	29.5%
Fine converter slag	2.5%
Fine sinter return	47.4%
Acid works cinders	1.1%
No 3 precipitate	1.0%
Converter flue dust	5.0%
Fine quartz & oxide ore	13.5%

The plant was designed to make 150 tonnes of sinter per 24 hour day but it never achieved more than 100 tonnes. Eliminations of 80% of the arsenic, 75% of the zinc, 60% of the sulphur and 20% of the lead were achieved. Operating costs, at ptas 22 per tonne, were four times the estimated figure! The plant was scrapped in 1937 and replaced by a briquetting plant.

Briquetting Plant

The idea of briquetting copper precipitate as a means of reducing losses when adding it the converters or smelting it in a rotary furnace must have originated in the London office. In 1937, a briquetting press was purchased from Bruck Kretsher of Osnabruck, Germany - it was designed for a dual purpose: to make briquettes either 80 mm diameter and 65 mm high from precipitate or 110 mm in diameter and 80 mm high from concentrate. A substantial tonnage of copper concentrates had been impounded in ponds in 1931-32.

The briquetting plant was built on the site of the sintering plant. It proved ineffective on precipitate but worked reasonably well on concentrate. When the concentrator restarted in 1942 one press proved inadequate and so did the single Dorrco filter of the original sintering plant. Although another filter was purchased, it was not possible to buy a second press from Germany in the middle of the Second World War and a copy of the original was successfully constructed in the RTC workshops at Huelva.

A small coke fired lime kiln was built at site, large enough only to supply the needs of the briquetting plant which required fresh lime; lime for the concentrator was brought from Niebla where it was produced in wood fired kilns; limestone for the briquetting plant kiln also came from Niebla.

Each briquetting press consisted of twelve moulds on a rotating table 1.75 m in diameter, with each mould charged in turn with the feed mixture. Upper and lower plungers compressed the charge at 300 kg per square centimetre for a period of 2.5 seconds; the compressed briquettes were discharged on to the table by the lower plunger.

The feed mixture consisted of about 86% concentrates, 7% flue dust and 75% lime. It was mixed in a pug mill and "conditioned" in a storage pit for about 24 hours; in this time the heat generated by slaking of the lime reduced the moisture content to about 6%.

Each machine produced about 35 tonnes of briquettes per 8 hour shift, equivalent to 1,000 briquettes, of 4.7 kilos each, per hour. The briquettes were transferred on platform waggons to open sided sheds and stacked to a height of about 2 m. They were left to harden for a minimum of 24 hours; longer in winter.

Although the briquettes were never as hard and strong as the smelter wished, the plant worked continuously from 1942 until 1965. Maintenance costs were high, mainly because of the wear on the pressure pistons and the inserted sleeves in the dies.

Herbert Potts, superintendent of the smelter for many years until his retirement in 1948, wrote:

> "The briquetting plant is a costly "White Elephant" and introduces a superfluous intermediary process; if it could be eliminated it would represent a very worthwhile economy."

Smelting After 1932

Although 6 blast furnaces were originally installed in the Bessemer Smelter, the average number in blast during the pyritic smelting period was 4.5 and it dropped to 3 when working with a semi-pyritic burden. By 1931, only one furnace was required with another maintained as spare. The "ordinary" furnace, as it came to be called to distinguish it from the Orkla furnaces (described in Chapter 14), smelted any cupreous material that was available; for example, cupreous pyrites, sinter, briquettes, copper precipitate and low grade Orkla matte. Table 2 shows the average burden smelted over the period 1939-57 and it will be appreciated that such average figures conceal wide variations.

The Momoda Process

In 1960, after RTC had transferred management to CEMRT, news of a process developed by Sumitomo Metal Mining in Japan attracted attention at Rio Tinto and, after a visit to the Japanese smelter and a meeting with the inventor (Mr Ryochi Momoda) in July 1963, a successful trial was made at Sumitomo's Shisakajima smelter on concentrates of the type then being smelted at Rio Tinto.

Two blast furnaces were altered to the Momoda design, with a centrally placed hopper through which the burden and coke were fed in alternate layers, and the first was blown in on 12 December 1965.

The Momoda process proved a significant improvement on the "ordinary" furnace operation - it was seriously considered for a new smelter of 40,000 tonnes pa capacity at Huelva; when the decision was made to build an even larger smelter the Outokumpu process was chosen.

That new smelter came into operation in 1970 and smelting at the mines ceased in April of that year.

TABLE 2 AVERAGE BURDEN 1939-57

	Tons/day	% on burden	Cu %	Tons copper
In				
Cupreous pyrites	39.1	10.35	2.42	0.95
Briquettes	77.3	20.47	12.82	9.91
No 3 precipitate	5.8	1.54	60.76	3.52
Orkla matte	108.0	28.59	7.56	8.16
Converter slag	66.1	17.50	6.49	4.29
Flue dust	0.4	0.10	15.79	0.06
Silica flux	69.7	18.45		
Limestone	11.3	3.00		
Burden	377.7		7.1	26.89
Coke	28.7	7.6		
Out				
Matte	101.6		23.84	24.22
Slag	224.2		0.4	0.92
Flue dust	13.6		8.91	1.21
Unaccounted				0.47

Some General Observations

The complex mineralisation of the Rio Tinto ores leads to the presence of minor elements in the copper produced and the problems of arsenic have been referred to at several points in these notes. Antimony is also an important impurity although it was rarely reported in RTC analyses of precipitate; at the Port Talbot smelter, the antimony content of fire refined copper varied from 0.018 to 0.028%. This low level did not adversely affect mechanical properties of the metal.

Rua Figueroa referred to the high antimony content of the black copper produced; in his time at the mines, mid-19th century, it was always reported as higher than the arsenic content and he quotes an analysis of 94.22% Cu, 2.0% Sb and 0.69% As. This may have been due to the difficulty of separating arsenic and antimony in the assay. In recent times, the antimony content of pyrites at Rio Tinto has varied between 0.005 and 0.06%.

Gold and silver are important impurities in Rio Tinto copper. In early operations with reverberatory smelting of precipitate it was established that gold, silver and some other impurities concentrated in the first metallic copper separating from the white metal. When the Bessemer smelter came into operation, selective converting was investigated and referred to by Collins (31) in 1885.

It was not, however, practised until 1934-35. Normally blister copper was shipped to the RTC smelter at Port Talbot for fire refining but at that time, and presumably for political/commercial reasons, RTC sold blister copper to two Spanish refineries, only one of which paid for gold and silver.

A method was developed converting 9 ladles of matte to make about 15 tonnes of white metal; slag was removed and blowing continued until there were signs of copper on the punch rods. Blowing continued for a further 5 minutes before the impoverished white metal was poured off slowly into ladles and added to another hot converter, perhaps with the addition of one or two ladles of fresh matte - the enriched copper in the first converter was blown for a few minutes and then cast.

Normal blister at this time contained about 17 gm Au and 670 gm Ag per tonne - using selective converting the gold content of the enriched blister could be raised above 80 gm. Between June 1934 and December 1935, 375 selective blows were made and 40% of these blows produced copper of more than 80 gm per tonne Au.

Smelting of the fine No 3 precipitate was always troublesome. At one time a cupola was tried but without success and the best method seems to have been the addition of dried "cakes" to the converters. Adding precipitate to the blast furnaces gave recoveries of only 88-89%, whereas converter smelting gave 92%.

In 1932, when a large stock of No 3 precipitate had been accumulated, one of the converters was altered; the mouth was removed and replaced by a domed roof, with an exit placed centrally in the roof for the flue gases. Thus modified it smelted some 7,500 kg of precipitate daily, using 1960 kg of oil. Although inefficient, this operation reduced the stockpile over the next two years.

About this time a trial at Port Talbot smelting No 3 precipitate in a reverberatory furnace gave a recovery of only 77.5%, with slag containing 14.5% Cu and unaccounted losses of 17%. the latter possibly due, at least in part,to difficulties in sampling the precipitate.

Typical analyses of blister copper produced in the Fundicion Pirita era were as follows:

	1934	1964
Cu %	99.01	99.30
As "	0.165	0.03
Sb "	0.020	0 012
Bi "	0.0012	0.0003
Pb "	0.120	0.097
Fe "	0.080	0.025
Zn "	0.008	0.025
Ni "	0.038	0.064
Co "	0.0015	0.005
Se "	0.0012	0.015
Ag gm p t	669	812
Au " " "	17	13
S about 0.06% throughout		

Note: 1934 operation was on Rio Tinto ores only; in 1964 custom concentrates were smelted.

Copper Production, RTC Smelters

From the start of the RTC operations the output of the smelters it established was as follows:

Fundicion Mina 1879-89

48,629 tonnes matte containing 31.80% Cu.

Fundicion Huerta Romana 1884-1901

236,638 tonnes matte containing 32.98% Cu.

Fundicion Bessemer 1901-14

164,312 tonnes blister containing about 98% Cu.

Fundicion Pirita 1907- 70

190,714 tonnes matte sent to Fundicion Bessemer 1907-14

550,673 tonnes blister containing about 99% Cu 1914-70

15,205 tonnes black copper containing about 97.5% Cu 1936-44

Figure 22 Bessemer Smelter, Marasmilla Dam and Cerda Leaching Area

CHAPTER 14 Sulphur production — the Orkla process

Although the Rio Tinto mines are perhaps most widely associated with copper production, it should be borne in mind that it was the demand for sulphur that stimulated the major development of the mines of the Iberian Pyrites Belt and the search for means of maximising the return from the sulphur content of the ore has never been far from the minds of those operating there.

Elemental sulphur is a more attractive starting point for producing sulphuric acid in the chemical industry and many processes have been proposed for its production from mineral sulphides. The decomposition of pyrites into pyrrhotite and elemental sulphur on heating - the separation of the labile sulphur atom - has long been recognised as the potential basis for a process; as recorded on page 62, experiments were carried out and six furnaces built in the mid-1880s but that attempt failed.

For a time efforts were directed to possibilities of a process based on the Claus reaction of sulphur dioxide and hydrogen sulphide - the idea was to produce a mixture of the two gases by passing steam over heated pyrites. Among other potential obstacles to the development of such a process, impurities in the pyrites introduced problems which, even if a technical solution could be found, would make the process hopelessly uneconomic.

Orkla Grube-Aktiebolag, operating a pyrites mine in Norway were similarly interested in recovering elemental sulphur and by 1929 had demonstrated a process in which cupreous pyrites is smelted in an enclosed blast furnace, designed to prevent ingress of air at the top, to recover both matte and elemental sulphur. Pyrites, fluxes and coke are fed through double bells, similar to those used on the iron blast furnace. Labile sulphur distills off as the charge descends and sufficient coke is charged to reduce some of the sulphur dioxide with hot carbon. Hot gases leaving the furnace are cleaned by electrostatic precipitators and cooled to condense the sulphur.

RTC had followed the Orkla development closely and ore from Rio Tinto was smelted at the installation at Orkla Grube's Lokken mine, with successful results. However, the sulphur produced contained a substantial amount of arsenic - Lokken pyrites contained less than 0.1% As; Rio Tinto pyrites more than 0.5% and arsenic content was no problem with sulphur produced from the former.

Following the successful smelting trial at Lokken, RTC decided to proceed with a new smelter to use the process. Wisely, before that major commitment was entered into, one of the furnaces of the existing smelter was modified for extended trials which began in August 1930.

No 6 furnace was used; the height of the shaft was increased to get effective heat transfer and reduce the off-gas temperature to below 460 degrees C. Four charging bells were fitted and two sets of electrostatic filters installed, with a cooling stage between to condense sulphur. The first gas filtration stage was to remove impurities such as lead, zinc, arsenic, iron and silica and the second to remove uncondensed sulphur mist before discharging gas to atmosphere.

Early operation encountered many problems. The furnace settler rapidly filled with magnetite and had to be dug out frequently. High arsenical sulphur was deposited in the electrostatic precipitators and mixed with the dust which was also deposited making removal extremely difficult. The plate type electrodes of mild steel were badly corroded and the quartz insulators failed.

It had been recognised following the Lokken trials that arsenic in the sulphur gave rise to two related problems - it was an impurity which could render the sulphur unsaleable and it significantly increased the viscosity of the molten sulphur. It was soon established that the arsenic content could be lowered to around 5 ppm by stirring milk of lime with the molten sulphur; it was some time before the problems associated with increased viscosity were resolved.

In the early operation at Rio Tinto it was found that the gas temperature had to be raised above 400 degrees C in the first electrostatic precipitator to prevent deposition of sulphur there and this meant that viscous sulphur containing about 20% As then deposited in the exit pipes - it was certainly easier to remove it from there but problems with viscous sulphur plagued the operation.

Corrosion of the electrodes was overcome by replacing the mild steel with 18% chrome steel, and redesign of the quartz insulators overcame the initial problems with those.

A technique for removal of arsenic from the product sulphur had been devised in RTC's English laboratory and, as first applied, consisted of stirring milk of lime into sulphur in two steam heated tanks, each of about 20 tonnes capacity. The operation depended on heat from the steam coils evaporating the water introduced with the milk of lime and was carried out at ambient temperature.

The design time was 5 hours per batch treatment, but actual time required proved to be 24 hours, due largely to the poor heat transfer from the steam coils. Experiments at the mines using an autoclave indicated significant improvements could be achieved with a pressure process but the London management was not convinced until it was learned that Orkla had independently developed an autoclave process.

A batch autoclave process was installed, although the continuous process developed by Orkla offered advantages, particularly since in the batch process operated at Rio Tinto a stable emulsion was formed as a layer intermediate between the sulphur and the arsenical liquor, leading to complications in discharging the vessel.

The cause of the emulsion formation was traced to lead sulphide impurity in the crude sulphur and no such problem of emulsion formation was experienced at Orkla Grube's operations - Rio Tinto pyrites contained over 1% Pb, Orkla pyrites around 0.05%.

In 1931 only 200 tonnes of "washed", that is, dearsenified, sulphur was produced but this increased to 7,146 tonnes in 1932. No new smelter for the process was built, largely because of the difficulties the company faced in operating during the Spanish Civil War. In 1932 No 5 furnace was altered to the Orkla design and a third furnace, No 4, was similarly altered in 1942.

It is of interest to note that Orkla built a "greenfield site" plant at Thamshavn which used catalytic reduction of the sulphur dioxide in the off-gases, thereby achieving a much greater recovery of the feed sulphur. Catalytic reduction was never a possibility for Rio Tinto because of the poisoning of the catalyst by arsenic.

A detailed account of the operation of the Orkla process at Rio Tinto is provided by Potts and Lawford(37). Tables 3 and 4 give operating data for the Rio Tinto plant in 1935 and a comparison with the Orkla Thamshavn operation.

TABLE 3 ORKLA PLANT DATA FOR 1935

Smelting

In	Tons	Burden %	Cu %	S %	Tons Cu	Tons S
Pyrites	104,499	68.54	3.74	48.43	3,904	50,609
Silica flux	26,636	17.47				
Converter slag	8,863	5.82	4.66	1.47	413	130
Matte	184	0.12	9.24	25.00	17	46
Limestone	9,824	0.44				
Sulphur residues	2,450					
	-------				-----	------
	152,457				4,334	53,063
Coke	10,603	6.95				
Out						
Matte	24,534		16.07	24.48	3,943	6,005
Slag	81,450		0.48	2.90	391	2,362
Flue dust *	637		2.43	20.03	15	128
Crude sulphur	35,600			97.17		34,600
					-----	------
					4,349	43,095

* Flue dust contained 22.0% Pb

Sulphur Washing Plant

In	Tons	As %	Ash %	S %		Tons S
Crude sulphur	35,000	2.22	0.59	97.19		34,600
Quicklime	1,323					
Out						
Washed sulphur	28,889	0.0005	0.01			28,889
Sulphur residues	2,450			93.0		2,278
Filter cakes	2,408			65.0		1,565
Arsenical liquor	19,890 m^3			92.1 kg/m^3		1,832

						34,564

TABLE 4 COMPARISON BETWEEN THE SULPHUR DISTRIBUTION FOR THE ORKLA FURNACES AT THAMSHAVN IN 1949 AND RIO TINTO IN 1935

	Thamshavn		Rio Tinto	
In	Tons	Recovery %	Tons	Recovery %
Sulphur in pyrites	99,945		50,609	
Sulphur in converter slags, etc			176	
Sulphur in residues			2,278	
Total	99,945		53,063	
Out				
Sulphur from pyrites	66,945	66.98	28,889	54.44
Sulphur from matte	2,998	3.00		
Sulphur from hot catalyser	6,635	6.64		
Sulphur from cold catalyser	4,608	4.61		
Total sulphur recovered	81,184	81.23	28,889	54.44
Sulphur in matte	2,953	2.93	6,005	11.32
Sulphur in slag	5,284	5.29	2,362	4.45
Sulphur in dust	432	0.43	128	0.24
Sulphur in gases	11,258	11.26	8,761	16.51
Sulphur in arsenical liquor	446	0.45	1,832	3.45
Sulphur in residues			2,278	4.29
Sulphur in filter cakes			1,565	2.95
Total sulphur	101,556	101.61	51,280	97.65
Gain/[Loss]	1,612		[1,243]	

CHAPTER 15 Miscellaneous activities

Although the emphasis in exploitation of the Rio Tinto deposits was initially on silver, then on copper and, in recent times, on pyrites, there have been notable instances of working for other values. Some of these other activities are summarised here.

GOSSANS; IRON ORE, GOLD AND SILVER

Extensive gossan capping is a feature of the sulphide lodes and interest in gossan as a source of iron production seems certain to go back to prehistory. As reported on page 15, the Romans used iron tools; they were experienced in smelting the metal and can confidently be assumed to have established local production.

According to Williams(21), there are two types of iron ore. The gossans are predominant and initially covered large areas of the North and South Lodes, varying in depth from 10 to 40 metres. The other type is a "pseudo gossan" or bog iron ore, and the largest formation of this type is the capping on slate, the Mesa de los Pinos. [It was here that Sanz, in the late 18th century, planted the pine trees which so impressed the Franciscan brothers whose visit is referred to on page 23.]

David Forbes, the consulting mining engineer, recognised the potential value of the overlying gossan and recommended separate dumping in removing it in the course of RTC's early opencast operations. Earlier, Spanish engineers had suggested that it could be used to make sponge iron for use in cementation of copper.

The mineralogical composition of the gossan is complex and, unfortunately as far as use as iron ore is concerned, it contains arsenic and lead. A trial shipment was made in 1881 and between 1887 and 1893 a total of 2,806,000 tonnes was sold. Though containing about 60% Fe, it presumably sold at well below market price for high grade ore.

In the 1920s and early 1930s, some bog iron ore was mined from the Mesa de los Pinos, but again proved uncompetitive against ores of higher purity. 1,496,000 tonnes was shipped.

RTC made a thorough investigation of the possibilities of producing iron at the Mines for use in cementation of copper. In 1896, two 3,000 tonne parcels of ore, one gossan and the other bog iron ore, were sent to the steel company, Bolckow Vaughan of Middlesborough, for the production of pig iron. Limestone from Niebla was used for the trial.

Pig iron produced by the trial gave the following analyses, all figures in per cent:

	Pig Iron Source		
	Gossan	Bog Iron Ore	Bilbao
Fe	93.52	92.94	92.95
Graphite	3.49	4.61	3.42
Combined C	0.55	0.35	0.52
Si	0.67	0.77	2.23
As	1.21	0.11	-
P	0.08	0.19	0.05
S	0.11	0.23	-
Mn	-	-	0.75

The Rio Tinto ores assayed, in per cent, as follows:

	Gossan	Bog Iron Ore
Fe	54.11	54.51
Silica	6.00	12.5
As	0.74	0.15
P	0.025	0.07
S	0.58	0.30

The pig iron from gossan was unsuitable for use in the copper precipitating tanks because of its arsenic content; that from bog iron ore was acceptable for the later stages.

In 1897, moves to build an iron smelter began but were frustrated by difficulties in obtaining title to the mineral rights over the Niebla limestone deposit. This, combined with a failure to identify a suitable coal deposit, following exploration at Villanueva de las Minas (to the east of Rio Tinto and 40 km from Seville) led to the project being abandoned.

As far as the activities of the Ancients are concerned, no slags have been positively identified as from iron smelting, but the vast accumulations of slag from silver and copper production could readily conceal the small quantities of slags from making iron for tools. An obvious place for iron smelting would have been along the Salomon ridge, where both brownish coloured and black slags were reported in early surveys.

Slags found in this area have generally been assumed to be from silver production. Smelting of gossan would produce a slag similar in chemical composition but would require a much higher temperature. An alternative possibility is that bog iron ore was smelted but no slags have been found in the vicinity of the Mesa de los Pinos. A third possibilty is that iron was produced by the Romans at another site and there is evidence of iron smelting at a site a few kilometres to the north.

The gossan contains small amounts of gold and silver. After the copper concentrator was closed down in 1932 (page 75), consideration was given to gold recovery using that plant. The Edquist process - cyanidation with gold adsorbed on activated charcoal recovered by flotation - was adopted. This required only modest investment in additional plant, avoiding the requirement for liquid-solid separation associated with the conventional "all sliming" cyanidation process.

Between 1937 and 1941, just over one million tonnes of gossan averaging 3.21 gpt Au and 49 gpt Ag were treated. Some 3,700 tonnes of concentrate, obtained by calcining the flotation concentrate, and assaying about 550 gpt Au and 2,400 gpt Ag, was produced and sold. Gold recovery was only 60.56% and silver recovery 17.3%; operating difficulties were considerable with a then novel process under Civil War conditions.

In recent years, RTC's successors have embarked on gold and silver recovery on a large scale. Their operations confirm the difficulty of silver recovery, possibly because it occurs in jarosites not amenable to cyanide dissolution.

PHYSICAL CONCENTRATION

Most of the production of the mines at Rio Tinto to the end of the RTC era was in the form of direct shipping or direct smelting mineral. One exception has been mentioned on page 74 - the concentration of copper from the cupreous stockwork. The second was the Lavadora plant for upgrading low grade pyrites to marketable level.

As described on page 63, the ferric sulphate leaching process used in the 1890s led to an accumulation of a mixture of partly roasted, partly leached mineral, known as morrongos. It was estimated that about 20 million tonnes had accumulated and represented a substantial potential source of pyrites "smalls".

In 1910, as the demand for smalls continued to grow, a pilot plant was built to separate the pyrites from the cinders, using Hancock jigs and a No 5 Wilfley table.

The success of the pilot plant in producing 48% S concentrate led to the building of the Lavadora gravity concentrator. It was located on the side of the hill to the north of Zarandas Naya, on a steeply sloping site and embodied the design philosophy then popular of gravity flow. It was equipped with Wilfley tables and jigs and underwent various modifications before shutting down in 1935.

Between 1912 and 1918 it treated 2,434,000 tonnes of morrongos to produce 1,039,000 tonnes of pyrites containing 42.5% S. By 1918, the impact of the First World War had led to a stock of 200,000 tonnes of product and the plant closed. It operated occasionally thereafter but was not a great success.

In 1954, a time of revived demand for pyrites because of brimstone shortage, Lavadora was rehabilitated using heavy media separation followed by flotation; all installed in the original building. The plant proved difficult in operation, the steeply sloping site was, on balance, a disadvantage, and the plant finally closed around 1960.

Testwork on concentration of the chloritic copper stockwork ore was done at Lavadora and the 1,000 tpd plant, built near the Fundicion Pirita, came into operation in 1930. It was of conventional design, with 2 stage crushing to about half-an-inch, single stage grinding in Hardinge mills to 80% minus 200 mesh, rougher flotation in pneumatic cells, followed by regrinding and differential flotation at high pH.

The high sulphur to copper ratio and close association of pyrite and chalcopyrite in the feed made it difficult to obtain high concentrate grade at high recovery. It was only by regrinding to 85 to 90% minus 325 mesh, that a recovery of approaching 90% could be obtained with a concentrate grade of 14 to 15% Cu.

The concentrator shut down in 1932 and restarted in 1942, when the Spanish government required maximum domestic production of copper.

Stockwork ore had been located in four main areas - South Lode, Salomon, Quebranta Huesos and San Dionisio. In the RTC era, most of the ore was mined from Salomon; the San Dionisio ore was not mined because of difficult access through the pyritic mass still to be mined. Pryor(38) has given a full account of the mining of the cupreous stockwork.

After the transfer of the Mines to Spanish control in 1956, exploration established a sufficient tonnage of copper porphyry to justify large scale open pit exploitation.

SULPHURIC ACID, SUPERPHOSPHATE

Initially, chemicals required by RTC were sent from the UK by ship - relatively small quantities were involved. In November 1883 the company was approached by a potential customer for sulphuric acid in Spain and, after taking a consultant's advice, told the potential customer in February 1884 that production would start in six months' time!

There were delays, chiefly due to difficulty in choosing plant suitable for the Mines, and it was not until January 1885 that discussions with Blaydon Manure Company of Newcastle-upon-Tyne led to purchase from them of design details for a fee of £500.

The lead chamber plant was operational by the end of 1889. It had 66 burners, taking pyrites screened -2.5 in + 1.0 in, assaying 4 to 5% Cu and 44.5% S. Cinders were smelted at Fundicion Huerta Romano. In 1891, 8,850 tonnes of pyrites was roasted and 13,125 tonnes of 51 Baume acid, equivalent to 8,400 tonnes of 100% acid, was made.

In 1902 a McDougall multi-hearth roaster was installed. By 1906, four McDougall roasters were in operation, taking half-inch pyrites, and the original lump burners were all scrapped. Fine cinders now went to the heaps for copper recovery by leaching.

By 1925 the condition of this acid plant had deteriorated to the point where it was decided to build a new one. This was again a lead chamber plant and it was commissioned in 1929. Sulphuric acid production became increasingly inefficient through the Civil War. RTC sold its interest in acid production in 1941.

Just prior to transferring control of the mines to Spanish interests, RTC had embarked on plans to build a sulphuric acid plant to utilise the off-gases from the Orkla smelter. These plans were brought to fruition with the commissioning of a contact plant, built by Chemiebau of Germany, in 1960. That was a sophisticated plant, designed to accept dirty gases of low and fluctuating strength, but it still encountered operating difficulties and did not reach design performance before the Orkla furnaces were shut down in 1965.

Around 1900, RTC was interested in the possibilities of superphosphate production and William Nash, head of RTC's Lands Department, made what we would now call a market survey. He reported that farmers in the region could be expected to use chemical manures from a reliable supplier. In 1898, a German firm had begun purchasing acid for their small artificial fertilizer plant in Huelva. The attractions of this outlet led to RTC purchasing the operation in 1906. It was twice enlarged, but was still a rather small operation, producing 240,000 tonnes of superphosphate in a ten year period, 1931 t0 1940.

In 1941, when RTC was short of local funds, the fertiliser plant was sold, together with the acid plant at the Mines. The latter operated until 1960.

COPPER SULPHATE, FERROUS SULPHATE

About 1882, Bordeaux vignerons discovered that copper containing sprays provided protection against the mildew which had recently become a serious problem in their vineyards. Production of copper sulphate was an attractive commercial opportunity for RTC and was actively pursued.

At first it was thought that copper precipitate (cement copper) could be converted to a suitable sulphate crystal but the product was too high in iron. Two reverberatory furnaces were installed and No 2 precipitate was refined and granulated to copper shot. This was partially oxidised in a small furnace, then dissolved in hot sulphuric acid, with air sparging. The hot solution passed to crystallizing pans, from which the mother liquor returned to the precipitating tanks.

The process was developed at RTC's smelter at Cwm Avon in Wales and operation of it commenced at the Mines late in 1889. Production reached a maximum of 5,889 tonnes in 1903 and then declined, presumably with competition from producers based on copper scrap closer to the consumers. The decline in copper sulphate production, which had provided the main offtake for sulphuric acid, influenced the decision to enter superphosphate production.

In the 16th century, and probably earlier, copperas had been collected from the banks of the Tinto river. In recent times the quantities of iron discharged to the river increased enormously but it was not until the 1930s that a serious interest in recovery of ferrous sulphate developed. At that time it was found that the soils of the Canary Islands are deficient in iron; Spanish producers supplying fertilisers there required a suitable means of adding iron to their product.

A simple method of production, based on solar evaporation, was used and the product was impure; it contained a few percent of ferric sulphate, more than 1% Zn and approaching 0.1% Cu. Such impurities were apparently of no great concern in the fertiliser additive; the copper is significant in relation to use in the Patio process referred to on page 18 and the zinc content was certainly the subject of attention about 1930.

At that time, attempts were made to recover zinc from the effluent from the precipitating tanks, where the level was about 2 gpl. The company had considerable experience with hydrogen sulphide in precipitating copper and extended the work to strip both copper and zinc, separately, from the effluent. While technically successful for zinc, because the sulphide precipitate was readily filterable, precipitated copper sulphide was extremely fine and difficult to filter and the process development was abandoned.

Appendix 1

ROMAN ACTIVITY ELSEWHERE IN THE IBERIAN PENINSULA

The activities of the Romans in others mining areas of the Peninsula have been investigated by individuals and by official bodies.

A Commission of Mining Engineers in Portugal was set up in 1865 under the Chief Inspector of Mines, Joso Maria Leitao, to consider activity in the Algares area - the following is a translated excerpt from that Commission's report:

> "The Algares mass does not outcrop as does that of San Joso do Deserto, nor is it accessible by either ancient or modern work. The contact at the western end is seen, but that at the east was not found, being covered with slag. The Algares valley is largely covered ... by numerous slag heaps, in which copper is visible as carbonates. The average assay of five samples of porous slag heavily charged with copper carbonates is 3.85% Cu. Flaky, lamellar slag contains as much as 3.86% Cu but the average of twelve samples is about 2%; this is not the average of the dumps because the most abundant slags contain only a few "millesimal" parts of copper. Finally, a compact slag did not reach 0.4% Cu."

The Commission thought that the last mentioned slag was attributable to the Romans and the richer slags were assumed to have resulted from operations by the Phoenicians. The Commission did not attempt an estimate of the quantity of slag, contenting themselves with a statement to the effect that it must amount to some hundreds of thousands of tonnes.

An assay of a sample of slag collected in the valley in 1960 gave 0.6% Cu and 3.0 gms per tonne silver, not markedly different from copper smelting slags at Rio Tinto.

It was in the Algares valley that two bronze tablets were found buried in the slag heaps; they are known as the Aljustrel Tablets, after the name of the mine of recent times. The first, 760 mm long by 500mm wide and 9.5 mm thick was found in 1876 and is now in the Museum of the Government Geological Services, Lisbon. The second was found in 1906 and is in the Ethnological Museum, Belem, Lisbon.

Both tablets refer to Roman mining law, but of different periods. The first relates to laws of the reign of Augustus, although dated well after his death; the second was inscribed in the reign of Hadrian (AD 117-138). In a lecture to the University of Durham Philosophical Society in February 1921 on "Some Aspects of Mining Laws under the Roman Empire", Henry Louis made extended reference to the 1876 tablet. The seventh clause on that tablet has usually been translated as "As to the Workers of Slag and Minerals" and states:

> "He who in the mining district of Vipasca, with the authority of the Procurator, may wish to smelt, clean, break up, screen and wash slags of silver and copper, dust from slags, and ore bought by weight and measured, as well as he who similarly takes up the carrying on of some work in the quarries, must declare before the lessee within three days the number of slaves and paid workmen going to be paid by him .... denarii per head on the last day of each month; and if they should not pay then, they must pay double.
>
> "He who may bring ores of copper and silver into the district of the mines from other places, must pay the lessee 1 denarius for every hundred pounds in weight.
>
> "The lessee has the right to seize any article as security, and if he is not paid what is owed to him, the lessee may confiscate any slag and any mineral that should be smelted, cleaned, broken up, screened, or washed, as well as any slabs of stone which may be found dressed in the quarries; those slaves and freedmen belonging to the copper and silver smelters who are working at the furnaces of their masters or patrons are exempted from this payment."

The Roman heading of the translated clause is "Scripturae scaurariorum et testariorum". Henry Louis thought the usual translation incorrect and that "testariorum" could apply to men working at the "test" or cupel in refining silver, an operation likely to have been carried out under state control. Other interpretations are that silver slag refers to litharge produced during cupellation, but the discovery of a highly siliceous grey slag, possibly Phoenician, suggests that the dressing of the slag probably refers to crushing to remove some of the particulate silica before resmelting with added iron oxide. Ancient copper slags probably carried copper as carbonate or metallic globules and would have been dressed to recover mineral and metal.

Henry Louis considered that the second tablet was the more important because it goes into more details about the laws governing the mining operations. There are two clauses referring specifically to silver mining shafts. One translates: "Whosoever out of the number of five shafts shall have sunk one down to the ore shall work without intermission in the others as is written above; unless he shall do so others shall have the power of occupying the same." The other: "Whosoever works silver shafts shall keep away from the drift and shall leave not not less than 60 feet on either side thereof." The drift would be the drainage tunnel carrying water out through a lower level of the mine. This suggests that the silver ore was above that of copper and damage to the drainage drifts would have adversely affected the mining of copper ore.

Mineralisation at Aljustrel includes veins of galena and tetrahedrite. When the mines were visited by Leitao, copper-rich slags, such as we now know to have been produced in smelting to recover silver, were to be found in substantial tonnage; these slags were reworked in the second half of last century for copper recovery.

Appendix 2

TRANSLATION OF REPORT BY DIEGO DELGADO

"Written report in the town of Aracena, 15 August 1556, by me, Diego Delgado, priest from Madrid, of the veins which we were commissioned to survey and check by Don Francisco de Mendoza and with Pedro de Aguilar, of Castro Nuno within the district of Zalamea la Vieja and other districts. In Zalamea, Don Francisco de Mendoza had seen large workings and buildings and equipment and tunnels and shafts which had been worked in ancient times and large slag heaps from the old foundries and workings. Having seen the layout of the area and of the land and boundaries and having seen that there were many veins, I, Diego Delgado and Pedro Aguilar were commissioned to particularly note, examine, check, find and discover as many metalliferous veins as were worked and recovered by the ancient people. Having checked these metal ores, Sr Don Francisco de Mendoza will send the most useful for his Majesty's service.

"Therefore, in compliance with the above provision, we set forth at the end of July 1556 to the district of Zalamea and we lodged ourselves in some houses on a mountain called Nuestra Senora de Rio Tinto about a quarter of a league from the tunnels and shafts made by the ancients. The day after our arrival we went to one of the caves, known as Salitre, and we entered it. This cave measures 70 paces wide and over 80 paces long. Its height is like a church and it is arched and within this cave there are great cavities and ventilation channels which come out in many places on the top of the hill. We believe the height of some of these ventilation channels is more than 15 estados(1) and light for the workings was obtained through these ventilation shafts. Likewise, within the cave there are some shafts which go downwards. I went down one of these to discover the reason why it was made by the ancients and in this shaft found a vein covered by earth which had been put there by hand. I cleared away this earth and discovered the vein and several openings in it, and from this vein and several holes I extracted an arroba(2) of metal in the time it takes to say six creeds. I felt that this was where the ancients carried out their workings.

"Having left the shaft, we walked the entire length of the cave and saw that the ancients followed four or five different metals in it, all of which started from one mother vein which ran along the cave(3). We saw that we would not be able to discover the metal in the cave due to the large amount of earth and stone left in it and we had to try and find it in the mountains, although the ground was very overgrown. As it was 10 days since our arrival at the cave and district of Zalamea, the Feast Day of St Laurence (10 August), I, Diego Delgado, hired a man with a pick-axe and went to a place where some days previously I had noted signs of a different metal and there I ordered him to dig and after two or three strokes I discovered metal. I ordered him to dig deeper up to knee-height and discovered a great quantity of metal of which I ordered an arroba to be taken for Don Francisco: this vein is little more than the throw of a harquebus from the mountain. Having seen the metal which I discovered, we deduced that it was all of one kind which the ancients had worked both in the cave and in the shafts. It was all of one material which I discovered and this I reported to Don Francisco so that he could order it to be examined, so that he could determine what should be recovered.

"In this way, walking in the examined area and in other areas, we saw very large foundations and smelter buildings and slag heaps from the veins which the ancients worked and recovered. These slag heaps are so massive they appear to be large hills and mountains. Could these slag heaps which we saw up to two leagues long and two leagues wide without doubt be those we were told were more than eight leagues long?

"We also visited another cave which was full of water and from under which sprang a river said to be the Rio Tinto(4). The reason why it is known as the Rio Tinto is because it springs from vitriol, which in another part is known as copperas, used for ink. All the sides of the river are covered with copperas mainly during the month of August and therefore in all the district through which the river flows, teams of women and lasses and lads are sent, during the month of August, to collect this copperas with which they pay certain taxes which they owe to the Archbishop of Seville. No other councils or other persons are allowed to collect it at any time under pain of grievous penalty because it belongs to the Archbishop. Up to now they did not know why the river was stained until I told them and informed them that it was because it springs from vitriol, although there is another secret in this which I kept to myself and did not disclose to them. As they saw and see that this river flowed tinted it cannot be called anything else but Rio Tinto.

"No fish or other life existed in this river, neither do people or animals drink it, nor are its waters used for anything else. The water has a property that if someone with a living organism in his body should drink it, the organism will die and be expelled from the body. Another property which I told them about and of which they were anxious to hear was that if any person had diseased eyes and washed themselves in that water they would be healed. This I gave them as medical advice and they were very pleased, because it was later proved to be true by experience. It has another property that if you place iron in the water it dissolves in a few days. This I tested and proved myself. I took a live frog and threw it in the river and it died without being able to leave the water. In all this river, no sand or loose material can be found because all the stones are fixed and stuck together and if a loose stone is thrown in, it too is stuck to the bottom of the river after a few days like the other stones.

"Walking in search of these veins,we found two great hills in which we found other methods of working the veins and shafts, some of which shafts were working for veins and others giving light and others of which had been worked in a different manner and had no metal, because as we found these had been made to dewater the veins which were being worked. These shafts interconnect one another. We found more than fifty shafts in a very large hill(5) which were used for dewatering, we went into one of these for further information and it was seen to be about 15 estados deep. Next to this shaft at about eight paces another river had its source and produced sour water which I also called Rio Tinto. There were many other things which we did not see owing to the very thick undergrowth on the hills and because we were told about them.

"We also searched where the ancients had their refineries to see if we could discover something which would shed light on the way they extracted silver or some other metal, but we could not discover anything because the hills could not be reached. Returning to our lodgings we found at the very top of a hill(6) the signs of a building. We ordered the men to dig there and at rather more than the height of a man we found a certain amount of lead from which we deduced that the ancients had worked in lead and that their aim was to extract silver.

"Asking many old people who had heard about these ancient buildings, they replied that they had heard say that in ancient times Spain used to give the Romans certain silver and gold talents as taxes and that they were taken from there, but they knew nothing more than that.

"Having seen all the above and having obtained all the information I could, we returned to Aracena where Sr Don Francisco de Mendoza was staying, to give him an account and report in accordance with our commission. Having arrived there we gave him all the information we could, together with the metal which we received which we gave over to him, so that he could arrange their examination and analysis."

Notes:

1. An estado is the average height of a man.
2. An arroba is about 11 kilos.
3. This refers to the varying colours exhibited by the jarositic silver ores.
4. Assumed to be what became known as Cueva del Lago.
5. Assumed to be San Dionisio.
6 Assumed to be Cerro Salomon.

Appendix 3

REPORT OF 1727 BY ROBERT SHEE

"In the district of Zalamea la Real is the mountain and location known in these days as the "Castillo Viejo" where are still to be seen walls and ruins, remains of buildings created by the ancients, on its summit, and which appear once to have been a castle or a fortress. From its elevated situation it was difficult of approach and commanded a good view of the surrounding country below, and it must have been a place of some strength.

"From this hill, in two different places, on the north and south sides, exude streams of strongly discoloured and mineralized water, highly charged with foreign matter and appropriately named at their junction and continuation the "Rio Tinto". The same name is given to an "aldea" or village of some 80 houses, distance about 5 leagues from Aracena, 20 leagues from Guadalcanal, 12 from Sevilla and 2 from Zalamea la Real. This latter town is a place of some 1,000 houses, with plenty of good water and fertile soil, and having an excellent situation and climate. Its people are industrious and well-to-do, having a considerable business in pigs, hides, honey and wax.

"The village of Rio Tinto is somewhat contiguous to the hill referred to, which extends some 3,000 yards, at a considerable elevation above the country below. The south-western part, or slope, of it has been traversed by a road still to be seen or identified in part, here and there, undoubtedly built by the Carthaginians or Romans for the transport of the products of these mines to the port and place of embarkation named San Juan del Puerto, situated some 9 leagues distant from Rio Tinto.

"So much exploration and work was performed at one part of the "Cordillera" that the slag resulting from the smelting furnaces forms huge mounds, many of them being equal in proportions to small hills at and about the foot of the "Cordillera" and have received, or being the subject of, the greatest admiration, and have deeply impressed all those who have visited the place and seen them.

"This is verified and substantiated, apart

from what is indicated, if not proved, by the evidence of labour, subterranean and superficial, especially in the hills of "escoriales" and in the ruins of a rather considerable town, to which tradition has given the name of "Gran Bitania", to be found widely scattered on the lower slope of the north side of the hill, on which the tower or castle of Salomon in former times stood.

"Here and there remains of high columns of cut stone of an iron conglomerate, indications and remains of foundations and other parts still remaining as they fell centuries and ages ago, amply demonstrate, beyond any shadow of doubt, the size and importance of the place - that it was once a large, prosperous and important colony.

"The Saracens also worked in this, the north side of the mines, and they, being skilful in such labour, constructed an aqueduct on the southern side at a much lower level than that of the site selected for working by the ancients on the north side. By it they worked towards the centre of the mine, with the object of being able to work at lower levels in procuring a natural drainage, without having recourse to any artificial means; and although today we find large deposits of vitriolized waters in the interior of this mountain, I believe that, as all might be made to drain out without the use of artificial means, it can be easily and quickly done and in that manner, at small cost, we can discover all the old, grand and important interior workings and riches.

"It is evident that the obstruction to the free exit of the water from the interior of the mine by that drain or aqueduct - for that is its present condition - is a stone of very large proportions, which, in my opinion, has been maliciously placed at the mouth of the eleventh shaft or ventilator of the aqueduct, in consequence of which the exit from the interior parts is interrupted and there is no doubt but the Saracens, who were the last to work in these mines, finding themselves hard pressed and obliged to retire from the place, adopted this means of preventing the Spaniards from any easy adoption of the benefit of their own and earlier labours and explorations.

"The tradition is commonly current in this neighbourhood that in the various hollows and caves of this "hill", in its subterranean works, exists hidden treasure in great quantity, either in money or in metal; but what foundation or probability exists for such I cannot well understand, unless it be that because of the sudden exodus of the Saracens after a long and prosperous occupation of the place, they must have been compelled to bury and leave behind, hidden, a certain quantity of treasure.

"For better explanation as to the aqueduct, or drain, referred to, it will be proper for me to state, your Excellencies, that it is so large as to permit the easy entrance of a person to work in it and that from its aperture at the foot of the hill to its conclusion there are 22 shafts serving for light and ventilation, at regular distance from each other of twelve yards, thus the total length of the gallery is 264 yards. In the mouth of the eleventh ventilating shaft there is placed the very large stone already mentioned, which detains and arrests the free exit and drainage of the water from the interior and from the lodes. By it is formed a substantial lake of highly vitriolized water, very clear and translucent, so much so that the bottom can easily be distinguished, as well as the obstructing stone.

"I had the curiosity to measure the depth of the water and ascertained it to be about 12 estados. The highest of these ventilating shafts, that is the last of them going up the hill, is some 30 estados in depth, 15 estados being in water.

"Of what particular kind of metal these ores are most abundant in, whether gold or silver, it is not easy to ascertain, but all indications lead one to infer that gold is the principal item. But apart from indications so patent which I have derived and deduced from documents of undoubtedly high authority, are the notices given by a most eminent subject on the working and treatment of the metals obtained in Peru, and under some guidance and teaching they are today being worked in the two kingdoms of the Indies.

"He came to this kingdom with two of the most expert assistants that could be found, under an order of King Philip the fifth, and from the year 1648 until 1651 they were employed in the study and inspection of various mines. In that of Rio Tinto they were for some time employed, and after many and various experiences, discovered the method of separating the mineral. In the memorial that he wrote of what he had done, he maintained that His Majesty had in these works and in the "escoriales" immense treasure, having found but little silver, but a considerable part of gold; but as His Majesty wished to charge all the expenses in relation to the matter to the Treasury and remain interested in all products, nothing resulted, partly because of the inefficient administration and large costs and because there also existed at the same time the war with Portugal; the ultimate result and end being that this site and its abundant minerals remained forgotten.

"The metals which are herein indicated and referred to are not those as given or as extracted from these mines, but are those which had already been suffered a first treatment by the ancients and were by them left deposited and hidden, with the intention of returning again to treat and assay them at a convenient time and opportunity. They are composed of many metals and different kinds of minerals, abounding in antimony, arsenic, sulphur and iron; for which reason in order to abstract the richer parts contained in them, it will be necessary to prosecute the method of extraction no doubt ascertainable in the archives of the year alluded to of 1651 and there are many men of great skill able to extract and separate pure metal from impure, in different kingdoms of the north.

"I am unable to specify definitely and clearly what are or will be pure metals to be obtained from these mines, but from all that I have mentioned, the existence of the deposits of vitriolized water in the interior, the copious quantities of "vermellon" and "caparrosa", natural and fine, that are also much in evidence, and of which I send samples to your Excellencies; all, in my judgement, undoubtedly indicate the abundance of and superior value of the metals to be found there.

"At what period of antiquity and history, or under what particular government, the first attempts were made to extract and assay these minerals, I am quite unable to ascertain and certify, but, undoubtedly, operations have more or less continuously been carried on here for very many centuries; for the present I will simply take the evidence of pieces of money and medallions, casually found by "pastores", herdsmen and others in and about the ruins of that ancient city of Gran Bitania and of the old castle on the hill already referred to, which I have been able to acquire and hand to His Majesty the King. One is a piece of silver, a coin distinctly showing the bust and name of Augustus Caesar; and yet others, which from their form, make and similitude to others I judge to be of the Caesars Trojan and Tiberius, also of copper and gold. Another, again obtained from a stone sarcophagus, is very distinctly new, as though but recently coined, but, as the characters impressed upon it are in Arabic, I cannot pretend to determine to what period it may be ascribed.

"Although it is not at all necessary in this instance to determine or ascertain exactly how, when, or by whom these evidently ancient works were commenced and abandoned, I maintain that the art of extracting and smelting minerals and the drainage of the mines is obliged no whit to the ancients; on the contrary, within a certain period of genius, industry and application of the people of the different kingdoms have been materially developed and advanced by the ingenious discovery of so very many admirable mechanical appliances, instruments and furnaces for the saving of labour in mines, and working in minerals and meatals generally, which today are beyond the comprehension of very many artificers and miners who have not yet made use of them or had any practical acquaintance with their application.

"This Sirs, I maintain is all pertinent to the matter and very much in its favour and to your advantage, in which supposition I conclude, supplicating that you will continue to be stimulated to and determined on the prosecution of this grand work.
Providential signs are not wanting in your favour, as already a mine has been found abundant in "carbon de fierro", as to the quality of which expert opinion has been taken at the town of Guadalcanal and I assert that it does not yield in character or favourable conditions to any of the very famous mines in England.

"At Cazalla, too, a no less valuable mine of azoque [mercury] has been discovered, the use, importance and value of which, in furnaces and in the separation and assaying of metals in general, is so notorious that I am excused from further praising it.

"Your company is not to be at the cost or charge of the Government nor under the charge of the vassals; you are neither contractors nor lessees of royal revenue, nor will your agents cause trouble or disturbance to the population of the district affected by your operations; you do not pretend to form treasure palaces or castles at the cost of others or of the public; finally you do not propose to create certain offices that may become a burden or charge to the State revenues; on the contrary, animated by a valour and zeal most noble and profitable to your sovereign and his country in general, you employ your own capital and resources in the extraction from the bowels of the earth of a treasure and common benefit to all.

"From this industry that you seek to re-establish may be hoped and awaited the alleviation of the vassal, the help and succour of the necessitous and the consolation of the widow and the orphan; all with considerations which invite Spain in general to contribute to and promote the happy success of it, which may reasonably be anticipated under the beneficence of the Almighty Power, the favour and assistance of our renowned and pious monarchs, Philip and Isabella, the support of your illustrious company, and of the intelligent directorate. In this confidence I continue, desirous of other and greater occasion to be able to employ my poor zeal and talents for the common good.

"Signed D Roberto Shee
Madrid 17 March 1726"

Appendix 4

COPPER PRODUCTION 1829 - 1849

Year	1.	2.	3.	4.
1829	35.00	37.50		72.50
1830	93.00	79.00		172.00
1831	82.00	72.75		154.75
1832	272.75	55.50		328.25
1833	190.00	39.75		229.75
1834	132.25	31.00		163.25
1835	155.50	26.00		181.50
1836	177.25	30.25		207.50
1837	124.00	24.25		148.25
1838	136.25	39.50		175.75
1839	81.75	35.50		117.25
1840	128.50	70.50	18.00	216.50
1841	55.50	63.00	54.00	172.50
1842	66.50	70.50	116.50	253.50
1843	96.25	46.75	206.75	349.75
1844	49.50	50.50	244.50	344.50
1845	41.25	51.25	241.25	333.75
1846	13.00	50.00	234.00	297.00
1847	12.25	47.75	327.50	387.50
1848	4.50	49.50	373.50	427.50
1849	12.00	7.75	85.00	104.75
Totals	1959	978	1901	4838

Key:

1. From ore smelting - tonnes
2. From precipitation "
3. From vitriols "
4. Total "

Copper from vitriols probably includes copper from artificial cementation.

Appendix 5

COMMENTARIES RELATING TO THE REMISA ERA

1. Visit by Richard Ford

Richard Ford lived in Spain from 1830 to 1833 and subsequently published "A Handbook for Travellers in Spain" (London, John Murray, 1847) from which the following extract covering a trip to Rio Tinto from Seville is taken:

"Passing through Italica, the high road to Badajoz is continued to the Venta de Pajanosa, 4 leagues, and turn off to the left over a waste of Xaras, cistus and aromatic flowers given up to the bees and butterflies to Algarroba, 1 league, a small hamlet, where bait. Hence three leagues over similar country to a mountain village, Castillo de Las Guardas, so called from its Moorish "atalaya"; here sleep. 5 leagues over a lonely "dehesa" lead next day to Rio Tinto.

"The red naked sides of the copper mountain, La Corbeza Colorado, with clouds of smoke curling over dark pine woods, announce from afar these celebrated mines. The immediate approach to the hamlet is like that to a minor infernal region; the road is made of burnt ashes and "Escoriae", the walls are composed of lava-like dross, while haggard miners, with sallow faces and darkened dress, creep about, fit denizens of the place; a small green coppery stream winds under the bank of firs, and it is the "tinged river" from whence the village takes its name. This stream flows out of the bowels of the mountain, is supposed to be connected with some internal undiscovered ancient conduit; it is from this that the purest copper is obtained; iron bars are placed in wooden troughs, which are immersed in the waters, when a "cascara", or flake of metal, is deposited on it, which is knocked off; the bar is subjected to the same process until completely eaten away. The water is deadly poisonous; no animal or vegetable can live near it, and it stains and corrodes everything that it touches.

"The mines were perfectly well known to the ancients, whose shafts and galleries are constantly being discovered. The Romans and Moors appear chiefly to have worked on the North side of the hill; the enormous accumulations of "escoriales" show to what an extent they carried on operations; these old drosses are constantly used in the smelting, as from the imperfect methods of the ancients they are found to contain much unwanted copper.

"The village is built about a mile from the mines, and was raised by one Liberto Wolters, a Swede, to whom Philip V had granted a lease of the mines, which reverted to the Crown in 1783. It is principally occupied by miners; there is, however, a decent "posada"; the "empleados" and official people have a street to themselves. The view from above the church is striking; the town lies below with its stream and orange groves; to the left rises the ragged copper hill, wrapt in sulphureous wreaths of smoke; while to the right the magnificent flat fir bank, which supplies fuel to the furnaces. "La Mesa de los Pinos" is backed by a boundless extent of cistus-clad hills, rising one over another.

"A proper officer will conduct the traveller over the mines, and then follow the ore through every stage of the process, until it becomes pure copper. Entering the shaft you soon descend by a well or "pozo" down a ladder to an underground gallery: the heat increases with the depth, as there is no ventilation; at the bottom the thermometer stands at 80 degrees Fahrenheit and the miners, who drive iron wedges into the rock previously to blasting, work almost naked, and what few clothes they have on are perfectly drenched with perspiration; the scene is gloomy, the air close and poisonous, the twinkling flickers of the miners' lamps taper blue and unearthly; here and there figures, with lamps at their breasts, flit about like the tenants of the hell of Eblis and disappear by ladders into deeper depths. Melancholy is the sound of the pick of the solitary workman, who alone in his stone niche is hammering at his rocky prison like some confined demon, endeavouring to force his way to light and liberty.

"The copper is found in an iron pyrites, and yields about 5 per cent. The stalactites are very beautiful; for whenever the water trickles through the roof of the gallery, it forms icicles as it were of emeralds, and amethysts, but these bright colours oxidize in the open air and are soon changed to dun brown.

"When the "Zafra", or rough ore, is extracted, it is taken to the "Calcination" on the brow of the hill, and is there burnt three times in the open air; the sulphur is sublimated and passes off in a cloud of smoke; the rough metal, which looks like a sort of iron coke, is next carried to be smelted at houses placed near the stream, by whose water power the bellows are set in action. The metal is first mixed with equal parts of charcoal and "Escoriales", the ancient ones being preferred, and is then fused with "Brezo", a sort of fuel composed of cistus and rosemary. The iron flows away like lava, and the copper is precipitated into an iron pan or "copella" below. It is then refined in ovens or "Reverberos", and loses about a third of its weight; the scum and impurities as they rise to the surface are scraped off with a wooden hoe. The pure copper is then sent either to Seville, to the canon foundry, or to Segovia to be coined."

2. Observations by Fernando Bernaldez

Fernaldo Bernaldez worked at Rio Tinto after the Remisa lease was terminated. Reference has been made in Chapter 2 to his publication in the Madrid journal, Revista Minera in 1853, of observations on the possible origin of ancient slags. In the same article he provides an interesting account of metallurgical operations in the years immediately preceding his employment at Rio Tinto; from this account the information which follows has been garnered.

The cupreous pyrites was broken to nut size, slowly roasted, leached and the copper precipitated on pig iron. An analysis of precipitate, or cascara, is given as:

Cu	71.45%
Fe	5.47%
Graphite	14.60%
S,As,etc	10.48%

In comparison, cascara obtained by natural cementation from the mine drainage waters analysed:

Cu	94.68%
Fe	3.98%
Silica,As	1.54%

Both cascaras were washed before being sent to the smelting house and sludge recovered from the washings contained 45 to 50% Cu. Sludge from the precipitating tanks contained 8 to 10% Cu. Both sludges were very fine and called "papucha".

Several small smelters treated all the cascara and any high copper content pyrites, the latter after roasting twice. The blast furnaces were of the Catalan type, each 1.6 metres high, and a typical daily charge was:

Roman slag	980 kilos
Papucha	300 "
Refinery slag	110 "
Charcoal	740 "

The output was 245 to 270 kilos of black copper, assaying typically 81% Cu, 17% Fe and 1.5% As. Papucha might be replaced by roasted ore; refinery slag assayed 50 to 55% Cu. If only cascara was smelted the charge was :

Roman slag	1010 kilos
Cascara	505 "
Charcoal	550 "

and this yielded 225 kilos of black copper.

Copper obtained from smelting papucha and roasted ore had to be remelted twice before it could be refined, no doubt in order to oxidize impurities. Refining of cascara from natural cementation and black copper was carried out in either German refining hearths or reverberatory furnaces.

A German refining hearth could process a charge of 500 kilos of black copper or good cascara in 24 hours, producing 400 kilos of copper suitable for alloying, or 290 kilos of refined copper, and requiring 480 kilos of charcoal in the first case, and 620 in the second.

In the reverberatory furnace, 1800 to 2000 kilos of black copper and/or natural cementation cascara was refined daily, producing 1400 to 1460 kilos of refined copper and using between 675 and 790 kilos of logs.

From his study of the relative merits of the two refining methods, Bernaldez recommended the installation of a larger reverberatory furnace and that was done in due course.

Appendix 6

COPPER PRODUCTION 1849-72

tonnes

Year	Ore Mined	Refined Cu
1849	3,117	301
1850	5,407	373
1851	6,558	579
1852	6,296	672
1853	5,373	537
1854	8,588	722
1855	8,193	785
1856	8,256	743
1857	8,283	650
1858	9,826	674
1859	12,849	953
1860	15,702	817
1861	17,603	1,136
1862	16,653	1,255
1863	89,694	1,355
1864	74,234	1,016
1865	66,156	1,025
1866	62,342	1,135
1867	50,480	829
1868	52,036	1,123
1869	60,530	974
1870	67,075	1,012
1871	55,600	860
1872	62,220	864

Appendix 7

FORMATION AND CORPORATE EVOLUTION OF THE RIO TINTO COMPANY LTD

The successful bid to acquire the mines was 92,800,000 ptas plus 1,195,912 ptas for valuation of plant, buildings, mineral stocks, etc; this was equivalent to £3,850,000. The consortium of financiers who made the bid included:

Messrs Matheson & Co of London
The Deutsche National Bank of Bremen
Messrs Smith, Payne & Smith of London
Messrs Arthur Heywood Sons & Co of Liverpool
The Union Bank of Scotland of Glasgow

Their offer was accepted on 17 February 1873. The property was purchased in perpetuity and consisted of 1,659 hectares (4,100 acres) and included nearly all the houses in Rio Tinto pueblo. The first payment in cash of £442,680 was made on 5 April 1873 and nine further "pagares" or promissory notes were to be redeemed on anniversaries of this date up to 1892.

The Spanish government was, in the years immediately following the sale, in serious financial difficulty and asked the Company whether these future payments could be made immediately at an appropriate discount. RTC raised the necessary money by issuing bonds and redeemed all the promissory notes by the end of 1876.

The Company was registered with a capital of £2,250,000 on 29 March 1873 and a prospectus was issued on 8 July 1873 offering the public 200,000 shares of £10 each. In 1881 the share capital was increased to £3,250,000 by the issue of a further 100,000 shares of £10 at £22 each. Then, in 1897, at the request of the larger shareholders, the Bearer shares were split into Preferred and Ordinary shares. The shares were split to £5 nominal value, after which the capital consisted of 325,000 Preferred and 325,000 ordinary shares. The rate of dividend on the preferred shares was set at 5%. A further issue of 50,000 £5 ordinary shares was made in November 1905 at £65 each. In May 1929 an additional 50,000 £5 ordinary shares was issued at £50 each. Thus, by 1930 the share capital was:

325,000 £5 preferred shares		£1,625,000
425,000 £5 ordinary shares		£2,125,000
	Total	£3,750,000

Further 5% Mortgage Bonds were issued in 1880, 1884 and 1892 and the earlier bonds issued to redeem the promissory notes, were repaid. The bond position in 1895 was as follows:

Date of Bond	Original Value	Outstanding Value
1880	£2,500,000	£1,862,340
1884	£1,200,000	£1,001,200
1892	£ 600,000	£ 573,520
	£4,300,000	£3,437,060

During July 1895, Messrs N M Rothschild & Sons converted the Company's mortgage debt and in the same year issued 4% First Mortgage Bonds to the value of £3,600,000. This mortgage was subsequently redeemed by the premium obtained with the issue of ordinary shares in 1905.

Financing the large enterprise which the Company had planned was difficult - the railway to Huelva had to be built, a pier constructed, dams and workshops built and the mine developed and equipped. Finally, new housing was needed for expatriate staff and the expanded workforce. A group of working Directors under Hugh Matheson controlled the day to day affairs of the Company - it consisted of Heinrich Doetsch, Dyes of the Bremen Bank and Edwin Clark of Messrs Clark, Punchard & Co, the railway constructors.

There was widespread scepticism about the prospects of the Company fulfilling the claims made in the prospectus and, although the Chairman gave glowing reports of progress at the annual shareholders meetings, confidence among shareholders declined - one shareholder pointed out that more money had gone into the mines than many German States possessed! A well-known Belgian mining engineer, M Julian Derby was engaged to provide an independent assessment and his report did something to restore confidence.

At the AGM for 1879, a Mr Dundas said:

> "I and others would be greatly interested if we could be informed when we reasonably may expect some return for our outlay. Like many others, I am one of the original shareholders of the Company, the men who have borne the burden and the heat of the day. In the prospectus when the Company was formed, it was stated that if present prospects continued, the Company might reasonably be expected to pay 18 or 20%. This certainly was very tempting and induced many of us to become shareholders. We know very well that the Directors cannot control prices any more than they can control the weather [a reference to the Chairman's reports at AGMs that the shortage of rain at Rio Tinto had hindered production of copper!]. We have been exceedingly anxious to hear what we have now heard, viz that we have now turned the corner, and I, like others, am very pleased to see the Company making a profit. What future rise must take place in prices before they reach the point where we shall be paid 5% ? I am not greedy - I do not want 18 or 20% although I should very much like it."

Mr Dundas and other shareholders were promptly rewarded; for 1879 a 5% dividend was paid, next year 8%, then 14%, and the company progressed amazingly for many years until the 1930 slump. Little was earned from the mines in Spain after that although, owing to wise financing, useful profits for shareholders were earned from other sources, as recorded by Harvey(32).

At the AGM of 1890, when the first dividend was declared, a Mr G Hamilton said:

> "In reviewing the past of our Company, I have noted very carefully its progress and development during a period when disaster was overwhelming us on all hands, and looking back on all this to the care and attention manifest for our interests by the firm of Matheson & Co, who, if I am rightly informed, provided more than a million of money when it was wanted .... I think, therefore, I shall only be expressing the general feeling of this meeting in proposing a cordial vote of thanks to our Chairman and his firm."

Mr Matheson, in replying and acknowledging the compliment, said:

> "I can only say, gentlemen, it was a great trial to us, in common with yourselves, when year after year passed and heavy and

increasing expenditure was incurred without any apparent satisfactory result."

Later, in 1898, the shareholders wanted to vote £5,000 in excess of the Directors' ordinary emoluments "for the fine way in which they have managed our business". Mr Keswick was then Chairman and he expressed his gratitude but said that the Board could not see its way clear to accept the offer.

During the period that RTC controlled the mines there were 6 Chairmen, 12 General Managers and 2 Interim Managers.

Chairmen

H M Matheson	1873-98
J J Keswick	1898-1904
C W Fielding	1904-23
Viscount Milner	1923-25
Lord Geddes	1925-47
Lord Bessborough	1947-54

General Managers

M W Carr	1874-79
Trew Prebbles	1879-85
A Blechynden	1885-86
J Osborne	1886-89
W Rich	1889-1900
W A Carlyle	1900-06
N Kennedy *	1906-07
W J Browning	1908-27
P S Couldrey	1927-32
F W Cooper	1932-34
G W Gray *	1934-35
A Hall	1935-41
A T Gough	1941-50
C R Julian	1951-54

* Interim Managers

On 20 June 1954 negotiations were completed for the sale of two-thirds of the Spanish assets of RTC to a Spanish financial group for £7.67 million. The Spanish banks parties to these negotiations were the Bancos de Bilbao, de Viscaya, Central, Exterior and Urquijo, together with the Banco Espanol de Credito and the Banco Hispano Americano. A new company, Compania Espanola de Minas de Rio Tinto SA (CEMRT) was formed with the above as shareholders and with RTC holding one-third of the equity. In 1962 RTC merged with the Consolidated Zinc Corporation to form the Rio Tinto-Zinc Corporation Ltd (RTZ) which thereafter held the interest in CEMRT.

A new company, Rio Tinto Patino SA (RTP) was formed in 1966 to develop the Cerro Colorado deposit for both gold from the gossans and copper from porphyry. This company's shareholders were:

CEMRT	55%
Patino Group (Canadian)	40%
RTZ	5%

In 1970 CEMRT merged with the larger Spanish company, Union Explosivas SA, to form Union Explosivos Rio Tinto SA (ERT). In the late 1970s Patino withdrew from RTP which was renamed Rio Tinto Minera (RTM). RTM acquired ERTs metal mining interests, RTZ later invested more money in RTM and the shareholders today (1986) are:

ERT	51%
RTZ	49%

and RTZ no longer has any interest in ERT.

Appendix 8

ENVIRONMENTAL - LAND, FORESTRY AND FARMING

The area of land acquired by RTC was 1,900 hectares (4,700 acres) but it was mostly barren and, according to the Government inventory of 1870, contained only:

Pine trees	equivalent to	4,252 m^3	of wood
Oak	"	" 1,875 "	" "
Poplar, elm & acacia	"	134 "	" "

In Roman times it is believed the country was covered with oak forests but during Roman occupation large areas were felled to provide charcoal for smelting and domestic use. In 1556, Diego Delgado observed only very thick undergrowth on the hills, and in 1634 Rodrigo Caro observed the lack of vegetation.

When the Franciscan brothers visited the area in 1780 they remarked on the presence of evergreen oaks and a lack of pines and praised Sanz for planting, a few years earlier, large areas of pines and some oaks. In 1832 Ford (see Appendix 5) referred to the magnifiicent pines on La Mesa de los Pinos, which had presumably been planted by Sanz 60 years earlier.

In 1839 there were 440,000 pines on the lease area. When the Prieto process was introduced in 1841, Remisa did not hesitate to fell pines to provide fuel to roast ore and this was also the beginning of the activities creating major pollution. The Spanish Government operations after Remisa continued to use the Prieto Process, limited at that time to roasting 65,000 tonnes pa of pyrites by mine output and availability of water.

Once RTC began operations, the tonnage roasted soon increased to 1,000 tpd, nearby villages, notably Nerva, suffered badly and considerable damage to the naturally poor agricultural land extended over an area of 76,000 hectares, affecting some 1,670 farmers owning some 11,000 separate properties. Assessment of damage was extremely difficult and an arbitrary scale was fixed on the basis of distance from the burning teleras - for example, owners of land within a radius of 6.5 km recieved 5% annually of the value of the land in crop, between 6.5 and 9.5 km distant the compensation was 4% and beyond that distance the rate was negotiated between 0.5 and 3.0%. Of course there was the uncertainty associated with "smoke farming" and special difficulties in assessing compensation for tree damage.

From 1889 to 1907, inclusive, RTC paid 3.7 million pesetas compensation. To reduce continuing liability, the company bought land where appropriate and in this way increased its land holding to almost 13,700 hectares by 1954. By that time, a great proportion of the land was forest, with some farm and grazing areas - some 2,400 hectares of waste land, 29% more than the area purchased in 1873, was still used by the mining departments for overburden, pyrites heaps, cinders dumps and the railway.

As early as 1877 the company experimented with tree plantings of Pinus maritima at Frailes, a station on the railway to Huelva and 12 km from Rio Tinto. Eucalypts, which had been introduced to Spain in 1865 by a British Ambassador and planted in a Barcelona park, were introduced to Rio Tinto in 1908 by W G Nash, an Australian in charge of the land department, who obtained them from Granada and planted them near Zumajo dam. The variety introduced to Barcelona was Eucalyptus globulus, known then as "Arbol de fiebre" as it was thought to purify the air and prevent fever (malaria). Nash planted more eucalypts near El Valle in 1914 and in Bella Vista in 1921 - there they grew quickly and soon obscured the view across the sierras.

Walter Browning, who became general manager in 1908, realised the need to cultivate land and plant trees if the company was to demostrate reductions in the damaging effects of emissions and he sought advice from Government forestry experts and they suggested that Pinus pinea was the best tree to plant in the particular conditions of limited soil capping and generally acid conditions. In their view, eucalypts could be grown only in small areas where there was suitable soil. In 1908 Browning had 120 eucalypt seedlings planted on marshy land purchased by the company the previous year near Huelva and pine seeds and acorns were sown at the mines.

Since 1901 the company had employed a Dane, Viggo Poulson, to run a farm which it owned at Huelva. Browning brought him to the mines to advise on land cultivation there and Poulsen's advice was to employ a forester, not a farmer. As a result of that advice, Kai Hasse, a graduate of the University of Copenhagen was engaged and arrived at the mines in 1916. He intensified the planting of Pinus pinea, using a technique in which seeds were sown directly into holes made in barren rocky ground with a spiked hammer. Many young seedlings died but further sowings were made and the ultimate survival has produced hillsides today covered with mature trees.

Open air roasting of heaps ended in 1907 and pollution by sulphur dioxide was confined to the stack gases of the smelter, yet claims for smoke damage continued to be made and Browning sought advice from the Government agricultural department in Huelva. The investigation by its entomologists showed that with the reduction of the sulphur dioxide content in the atmosphere, many harmful insects had reappeared and were the cause of crop damage - they suggested that the situation would correct itself by natural processes and no further compensation was paid by the company.

By 1925 some 3,900 hectares had been planted with trees and 3,085 hectares were being used for pasturing goats, sheep and cattle and tree planting was halted in that year. Intensive planting of eucalypts started in 1920 and many varieties were introduced. Contrary to the earlier doubts expressed by Government foresters, eucalypts have grown well at the mines and are now widely grown throughout the province, supplying logs to a cellulose factory at San Juan del Puerto.

In 1935, as an economy measure, the company dispensed with the services of the forester and did not employ one again until 1949 and during that period the plantations suffered, from three main causes:

Much damage was done during the Civil War by fugitives leaving camp fires burning.

Contractors employed to fell trees for fuel wood cut down the biggest and strongest trees instead of thinning out the weak ones.

Trees were trimmed excessively by charcoal burners.

The land department was revived in 1949 and some 1,500 hectares were planted with eucalypts, chiefly Eucalyptus rostrara and Eucalyptus globulus, between 1950 and 1955.

Apart from tree planting, Browning and the RTC chairman of that time, Sir Charles Fielding, were interested in getting land cultivated. Although the farm at Huelva supplied milk and produce to the mines, production nearer to the site offered benefits and, with the cessation of heap roasting, encouragement was given to small farms in the mines area. The most important of these was at Zumajo but it was not until 1920 that a major effort was made to improve it. In that year a long strike affected the farm workers on the Huelva farm and Kai Hasse brought cows from it to Zumajo and commenced milk production in earnest. Thereafter it supplied milk, eggs, vegetables and fruit, although it was never a profitable enterprise.

Appendix 9

FUNDICION MINA

Furnace Charge 1880 - 1889

In	Tonnes	Cu %	Tonnes Cu
Calcined papucha	7,471	30.71	2,294.3
Dry raw papucha	215	20.25	43.5
Crude mineral	4,595	7.78	357.5
Crude quartz	28,800	2.02	581.7
Calcined mineral	170,549	6.61	11,273.3
Calcined quartz	30,753	3.72	1,144.0
Kernels	899	9.85	88.6
Precipitate	2,555	15.23	389.1
Copper slag	3,988	5.02	200.2
Burden	249,825	6.55	16,372.2
Coke	47,467	(= 19% on burden)	
Out			
Matte	48,629	31.87	15,494
Slag	148,779	0.48	714

Matte Produced & Exported From Huelva
1880 - 1889

	Tonnes	Cu %	Tonnes Cu
1885 matte	17,945	34.5	6,191
1885-88			
1st class matte	8,314	37.8	3,143
2nd class matte	10,942	32.7	3,578
1885-89			
3rd class matte	10,123	22.3	2,257
1885-87			
4th class matte	1,051	21.6	227
	48,375	31.8	15,396

In 1881, only 3,027 tonnes of burden was smelted - it contained 13.75% Cu and produced 1166 tonnes of 33.2% Cu matte. In 1888, 42,983 tonnes was smelted - it contained 6.24% Cu and produced 8,070 tonnes of 30.8% Cu matte plus 31,669 tonnes of 0.54% Cu slag.

Appendix 10

THE BESSEMER SMELTER AND ITS OPERATIONS 1901-14

Blast Furnaces

The blast furnace feed materials were delivered by rail at a level above the charging floor and dumped into stalls. The charge components - ore, limestone, silica flux, coke, etc. - were shovelled into steel wheelbarrows of 400 kg capacity. The wheelbarrows were weighed before dumping their contents in front of the furnace charging mouths; men then pushed the charge into the furnaces. The method of charging was seen as a big improvement over that used at Fundicion Huerta Romana, where men carried the charge in baskets on their heads and climbed a ladder to dump the charge into the top of the furnace.

Slag, overflowing from the settlers, ran directly into small, conical shaped, cast iron pots, carried on two-wheeled buggies. When filled, the buggies were pulled by three men to the slag dump alongside the river - much of the slag was adventitiously granulated and carried away from time to time when the river was in spate. The two-wheeled buggies were replaced by four-wheeled trucks running on narrow gauge rail track - two men could then handle the hauling and dumping.

Converters

Initially the converter shells were lined with a 4 in layer of firebricks after which wooden formers were fitted. A pugged mixture of silica and clay was shovelled between the bricks and formers and pressed into place by men using their hands and feet - holes for the tuyeres were made by 1 in dia rods acting as formers. After lining, the top sections (hoods) were bolted on and the linings dried out, first by firing with wood, then with coke.

When relining, the firebricks were not replaced and only occasional patching of the hoods was needed. The lower sections of the lining were badly attacked by molten iron oxide and relining was needed after producing only 10 tonnes of blister copper! This called for relining five shells daily and must have been a very demanding operation. Shells were removed by crane and rapidly quenched with water; silica was crushed and pugged with clay in a Chilean mill and each relining required 7,200 kg of silica and 500 kg of clay - siliceous copper ore was used as far as possible and clay was brought from the Seville area.

A newly lined converter would hold 3 tonnes of matte and its capacity increased to 8 tonnes immediately before relining. Charging the 45-50% Cu matte was by means of an 8 tonne capacity ladle; copper precipitate was added after the converter had been slagged.

In 1902, the first full year of operations, the blast furnaces smelted:

	Tonnes	Cu %	Tonnes Cu
Crude pyrites	27,719	5.54	1,480.2
Roasted pyrites	63,071	6.02	3,796.9
Rich slag	312	30.78	96.0
Matte	46	45.07	21.0
Copper precipitate	6,372	48.14	3,067.5
Flue dust	469	5.56	26.1
Burden	97,989	8.75	8,487.7
Coke	13,689 (= 14,1% on burden)		
Slag produced	87,547	0.68	595.3

The converters produced:

Blister copper	7,317.5	98.5	7,207.7
White metal	28.2	74.2	20.9
Flue dust	1,147.0	8.43	541.8

The analysis of blister copper for February of that year was:

Cu	98.7%
Sb	0,012%
As	0,035%
Bi	0.0026%
Au	7.7 gm per tonne
Ag	45.0 " " "

Three significant changes affected the Bessemer smelter:

1. A third blast furnace was installed for pyritic smelting.

2. Open-air roasting of cupreous pyrites ceased in 1907 because all mine output was required for the new Fundicion Pirita which came into operation that year.

3. In about 1911 the linings for the converters were changed from acid silica to basic magnesite, using bricks.

In 1908 the low grade furnace matte containing 16 to 18% Cu was first smelted in combination with the last of the roasted ores and the furnace charge was:

Roasted ore	6,712 tonnes
Crude ore	1,212 "
Roasted quartz	5,956 "
Crude quartz	16,696 "
No 3 precipitate	9,302 "
Converter slag	36,863 "
Low grade matte	41,106 "
Flue dust	16,872 "
Limestone	5,660 "
	140,379 tonnes
Coke	13,285 "
	(= 9.46% on burden)

The converters also melted 1,304 tonnes of No 1 precipitate and 74 tonnes of scrap copper, producing a total of 14,592 tonnes of blister, assaying 98.17% Cu, and 5,863 tonnes of slag. No 1 blast furnace worked for 6182 hours; No 2 for 4006 hours and No 3 for 5482 hours, for an average of 210 tonnes of burden smelted per furnace day. 1474 converter shells were lined, for an average of 9.9 tonnes of blister per lining!

Following the success of Pierce and Smith at the Baltimore Copper Smelting and Rolling Company with magnesite brick lining in 1909, RTC began to use such lining in 1911. For the efficient working of the basic lined converters, good quality silica flux was required, certainly equal to that used previously in the silica lining. Because there was a shortage of silica, the flux was made by mixing decomposed porphyry, containing 3 to 4% Cu, with diabase, in the proportions two to one. The free silica amounted to only 60% and the resultant slag was viscous, probably because of the alumina content of the diabase rock.

Appendix 11

PYRITIC SMELTING

Furnace Charge 1913

	Tonnes	% on burden
Cupreous pyrites	260,688	77.73
Quartz	30,685	9.15
Limestone	13,993	4.17
Clean matte	815	0.24
Converter slag from Bessemer smelter	28,185	8.41
Flue dust	538	0.16
No 3 precipitate	474	0.14
Burden	335,378	
Coke	11,285	3.36
Charge	346,663	
Output:		
Matte	45,328	21.51% Cu
Slag	203,615	0.42" "
Flue dust	14,079	

Pyrites feed contained between 3 and 4% Cu; quartz contained 3.63% Cu, 14.3% S.

Appendix 12

SUMMARY OF PRODUCTION AND MANPOWER DURING THE PERIOD OF RTC OPERATIONS 1874 - 1954

Year	Tons Pyrites Mined Sales	Local Treatment	Total	Cu %	Washed Ore	Tons Cu Produced at Mines	Employees
1874						900	950
1875						932	1,132
1876	189,982	159,196	349,178	1.5		946	3,169
1877	251,360	520,391	771,151	2.38		2,495	4,152
1878	218,818	652,289	871,107	2.78		4,184	4,452
1879	243,241	663,359	906,600	2.78		7,179	5,439
1880	277,590	637,567	915,157	2.87		8,559	7,126
1881	249,098	743,949	993,047	2.75		9,466	8,364
1882	259,924	688,307	948,231	2.89		9,740	11,851
1883	313,291	786,682	1,099,973	2.96		12,295	9,801
1884	312,028	1,057,890	1,369,918	3.23		12,668	9,003
1885	406,771	944,694	1,351,465	3.10		14,593	8,273
1886	336,548	1,011,833	1,378,381	3.05		15,863	9,017
1887	362,796	819,642	1,182,438	3.05		17,813	9,986
1888	434,316	969,317	1,403,633	2.95		18,522	9,233
1889	389,943	824,380	1,214,323	2.85		18,807	8,955
1890	396,949	865,405	1,261,274	2.88		19,183	9,050
1891	464,027	972,060	1,436,087	2.65		21,327	10,576
1892	406,912	995,151	1,402,063	2.82		20,017	10,260
1893	477,656	854,346	1,332,002	3.00		20,887	10,305
1894	498,540	888,555	1,387,095	3.03		20,606	10,288
1895	525,195	847,181	1,372,376	2.82		20,762	8,761
1896	591,752	845,580	1,437,332	2.93		20,817	8,717
1897	575,733	812,293	1,388,026	2.81		20,826	8,697
1898	644,518	820,882	1,465,380	2.85		20,426	8,744
1899	644,271	1,005,573	1,649,844	2.72		20,230	9,377
1900	704,803	1,189,701	1,894,504	2.74		21,120	10,243

Year	Tons Pyrites Mined Sales	Local Treatment	Total	Cu %	Washed Ore	Tons Cu Produced at Mines	Employees
1901	633,949	1,294,827	1,928,776	2.63		21,000	10,475
1902	627,967	1,237,322	1,865,289	2.53		21,659	10,104
1903	688,919	1,229,619	1,918,539	2.39	118,174	21,565	10,161
1904	672,344	1,276,475	1,948,819	2.34	157,810	21,218	10,447
1905	627,336	1,202,768	1,830,104	2.36	308,184	19,530	11,486
1906	655,328	1,268,388	1,923,716	2.41	477,843	21,287	13,108
1907	641,959	1,265,090	1,907,019	2.42	619,814	21,251	14,221
1908	604,275	1,115,610	1,719,885	2.27	668,477	24,256	16,465
1909	604,799	1,184,188	1,788,987	2.35	569,604	24,364	16,873
1910	637,020	1,509,745	2,146,705	2.10	683,605	22,790	15,740
1911	649,215	1,536,390	2,185,605	2.14	841,964	21,880	15,449
1912	698,399	1,708,570	2,406,969	2.18	977,812	25,623	15,399
1913	652,168	1,207,403	1,859,571	2.19	729,099	21,400	14,604
1914	606,252	636,570	1,204,822	2.13	637,348	17,544	13,709
1915	264,455	789,767	1,054,222	2.20	676,163	20,357	14,633
1916	279,999	1,248,709	1,528,708	1.98	741,387	21,385	13,869
1917	224,682	1,156,436	1,381,118	2.22	531,667	27,005	14,054
1918	141,979	1,481,030	1,623,009	1.95	205,062	23,244	12,459
1919	72,464	1,343,730	1,416,194	1.85	212,560	21,361	11,145
1920	59,464	863,384	922,848	1.59	182,182	16,294	10,396
1921	252,517	1,665,559	1,918,076	1.37	402,548	18,124	7,996
1922	560,253	1,145,198	1,705,451	1.28	502,993	23,397	8,108
1923	470,989	1,541,092	2,012,081	1.28	419,331	22,090	8,154
1924	706,794	1,731,995	2,438,789	1.21	529,605	26,132	8,676
1925	857,801	1,575,830	2,433,631	1.21	445,250	27,370	8,817
1926	879,125	1,657,369	2,536,494	-	328,131	24,751	9,090
1927	835,802	1,673,863	2,509,665	-	307,262	25,270	9,914

Year	Tons Pyrites Mined: Sales	Local Treatment	Total	Cu %	Washed Ore	Tons Cu Produced at Mines	Employees
1928	857,850	1,549,878	2,407,728	-	498,426	23,670	9,813
1929	981,268	1,566,062	2,547,330	1.79	515,495	25,237	10,486
1930	800,822	1,446,993	2,247,815	1.94	515,203	22,742	10,994
1931	558,098	1,242,789	1,800,887	2.04	413,525	22,838	10,410
1932	515,856	697,014	1,212,870	1.83	387,673	15,726	9,923
1933	701,417	567,778	1,269,195	1.66	431,765	15,695	9,739
1934	636,538	530,374	1,166,912	1.74	590,454	11,968	9,584
1935	679,542	454,733	1,133,275	1.91	480,291	12,833	8,304
1936	656,103	247,352	903,455	1.49	379,089	12,192	6,757
1937	898,499	255,355	1,153,804	1.43	346,744	9,354	6,629
1938	951,170	167,357	1,118,527	1.45	252,460	8,721	6,490
1939	686,222	141,848	828,070	1.37	266,181	6,667	6,848
1940	451,840	277,344	729,184	1.31	171,403	4,728	6,755
1941	181,884	165,049	346,933	1.43	111,663	4,650	5,252
1942	191,989	219,381	411,370	1.35	121,709	6,756	5,111
1943	265,292	248,186	513,478	1.12	75,567	7,808	5,579
1944	157,668	233,307	390,975	1.28	78,695	7,634	5,587
1945	302,869	162,244	465,113	1.15	96,290	4,606	5,118
1946	360,755	175,697	536,452	1.08	149,328	6,594	5,362
1947	476,774	118,839	595,613	0.99	89,359	5,742	6,570
1948	517,094	132,200	649,294	0.93	204,047	5,428	6,386
1949	526,886	145,690	672,576	0.89	190,551	5,720	6,752
1950	627,237	161,072	788,309	0.88	172,223	6,107	6,683
1951	676,889	149,050	825,939	0.90	281,947	5,145	6,944
1952	739,171	153,723	892,894	0.86	233,583	6,008	7,358
1953	576,537	173,165	749,702	1.00	142,669	6,182	7,513
1954	648,816	157,283	806,099	0.92	168,471	5,451	7,916

REFERENCES

1. Avery D. Not on Queen Victoria's birthday: the story of the Rio Tinto mines (London: Collins, 1974), 464p.

2. The mineral wealth of Huelva: being an illustrated article on the history of this district from the earliest times. Min. J., Lond., vol. 64, 1894, 153, 171, 198, 233, 255, 293, 311, 366, 401, 457, 478.

3. Allan J.C. Considerations on the antiquity of mining in the Iberian Peninsula. Occas. Pap. R. Anthropological Inst. no. 27, 1970, 42p.

4. Blanco A. Luzon J.M. and Ruiz D. Excavaciones arquelogicas en el Cerro Salomon (Riotinto, Huelva) (Sevilla: Universidad de Sevilla, 1970), 39 p., 22 plates, 382 figs.

5. Blanco A. and Luzon J.M. Pre-Roman silver mining at Rio Tinto. Antiquityu, vol. 43, June 1969, 124-31.

6. Courtney L.F. Mining and treatment of copper ore at Tharsis, Spain. Min. Proc. Instn civ. Engrs, vol. 125, 1896, 126-44.

7. Gonzalo y Tarin Joaquin. Descripcion fisica, geologica y minera de la Provincia de Huelva (Madrid: Comision del Mapa Geologico, Provincia de Huelva, 1888), 2 vols.

8. Luzon J M Tartessos y la ria de Huelva. Seminar Arqueologica Salamanca, 1962.

9. 2 Chronicles, 9,13 ; 1 Kings, 10,22.

10. Delgado Diego. Letter to King Philip II, 1557. (Held in Spanish Archives; reproduced in reference 15, page 118).

11. Caro Rodrigo Caro was a famous poet and archaeologist living in Seville in the first half of the 17th century; the source of this quotation in his many publications has not been established - Ed.

12. Strabo The geography of Strabo with an English translation by Horace Leonard Jones (Cambridge, Mass.: Harvard University, 1961-1969).

13. Anciola Antonio and de Cassio Eloy. Memoria sobre las minas de Rio Tinto (Madrid 1856).

14. Rua Figueroa Ramon. Minas de Rio Tinto: estudios sobre la explotacion y el beneficio de sus minerales (Coruna, 1868).

15. Rua Figueroa Ramon. Ensayos sobre la historia de las Minas de Rio Tinto (Madrid, 1859).

16. Thomas Joseph Lee. Notes on the mines of Rio Tinto (London: Warren Hall, 1865).

17. Allan J.F. The pyrites deposits of the province of Huelva. Trans. N. Engl. Inst. Min. mech. Engrs, vol. 37, 1887-88, 27-51.

18. Palmer R.E. Notes on some ancient mine equipments and systems. Trans. Instn Min. Metall., vol 36, 1926-27, 299-336.

19. Bernaldez Fernando. Metalurgia del cobre en Rio Tinto. Revista Minera, vol. IV, Madrid, 1853, 252-65.

20. Tylecote R.F. A history of metallurgy (London: The Metals Society, 1976), 29.

21. Williams D. The geology of the Rio Tinto mines, Spain. Trans. Instn Min. Metall., vol. 43, 1933-34, 593-678.

22. Rickard T.A. With the geologists in Spain: the Rio Tinto mines - Roman slags and ancient mining implements. Engng Min. J., vol. 124, July 2 1927, 16.

23. Nash W.G. The Rio Tinto mine, its history and romance (London: Simkin, Marshall, Hamilton, Kent & Co., 1904).

24. Percy J. Metallurgy: the art of extracting metals from their ores - Silver and gold-Part I (London: John Murray, 1880), 561.

25. Gonzalez. Registro de minas de la corona de Castilla (Madrid, 1832).

26. de la Escosura Luis De la fundicion de las minerales de las minas de Rio Tinto. Anales de Minas, vol. 3, 1844/45.

27. de Prado Casiano Minas de Rio Tinto: Memorias (Madrid, 1856).

28. Checkland S.G. The Mines of Tharsis: Roman, French and British enterprise in Spain(London: George Allen & Unwin, 1967), 288 p.

29. Barton D.B. The Cornish beam engine (Truro: D Bradford Barton, 1965), 285 p.

30. Julian C.R. Underground mining at Rio Tinto, Spain. Trans. Instn Min. Metall., vol 49, 1939-40, 237-312.

31. Collins J.H. On the geology of the Rio Tinto mines, with some general remarks on the pyritic region of the Sierra Morena. Q. Jl. geol. Soc. Lond., vol. 41, 1885, 245-65.

32. Harvey C.E. The Rio Tinto Company: an economic history of a leading international mining concern 1873-1954 (Penzance: Alison Hodge, 1981), 390 p.

33. Taylor J.H. and Whelan P.F. The leaching of cupreous pyrites and the precipitation of copper at Rio Tinto, Spain. Trans. Instn Min. Metall., vol. 52, 1942-43, 35-96.

34. Potts H.R. Present-day smelting practice at Rio Tinto, Spain. Trans. Instn Min. Metall., vol. 38, 1928-29, 421-30.

35. Potts H.R. Notes on converting low-grade matte at Rio Tinto. Trans. Instn Min. Metall., vol. 38, 1928-29, 431-49

36. Potts H.R. Further notes on converter practice at Rio Tinto Trans. Instn Min. Metall., vol. 57, 1947-48, 337-75.

37. Potts H.R. and Lawford E.G. Recovery of sulphur from smelter gases by the Orkla process at Rio Tinto. Trans. Instn Min. Metall., vol. 58, 1948-49, 427-462.

38. Pryor R.N. Mining of cupreous stockwork ores at Rio Tinto, Spain. Trans. Instn Min. Metall., vol. 69, 1959-60, 661-86.

39. Collins J.H. On the economic treatment of low-grade copper ores. Trans. Instn Min. Metall., vol. 2, 1893-94, 4-57;59-75.

INDEX

RIO TINTO MINES.
GENERAL PLAN OF WORKS.

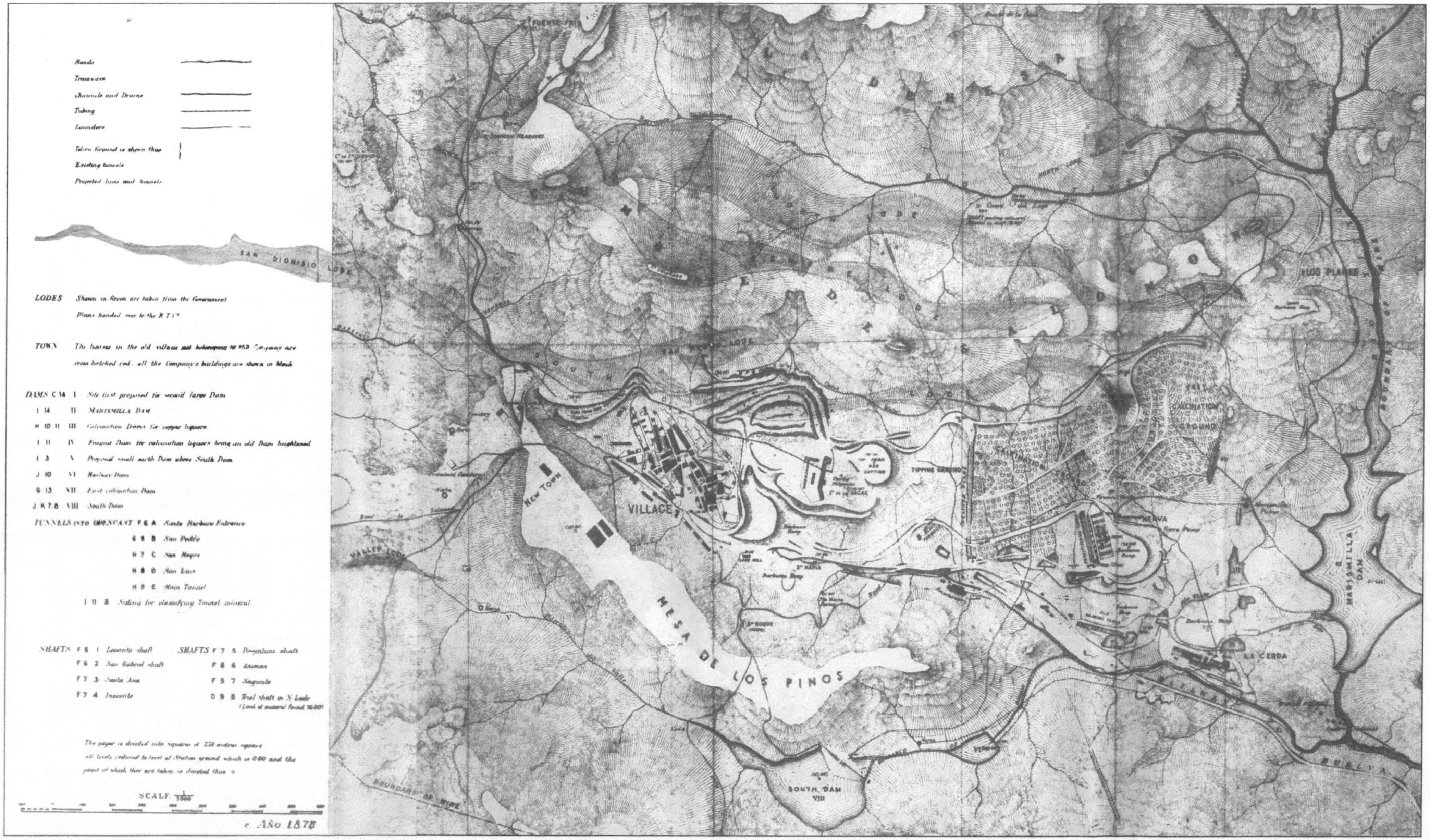

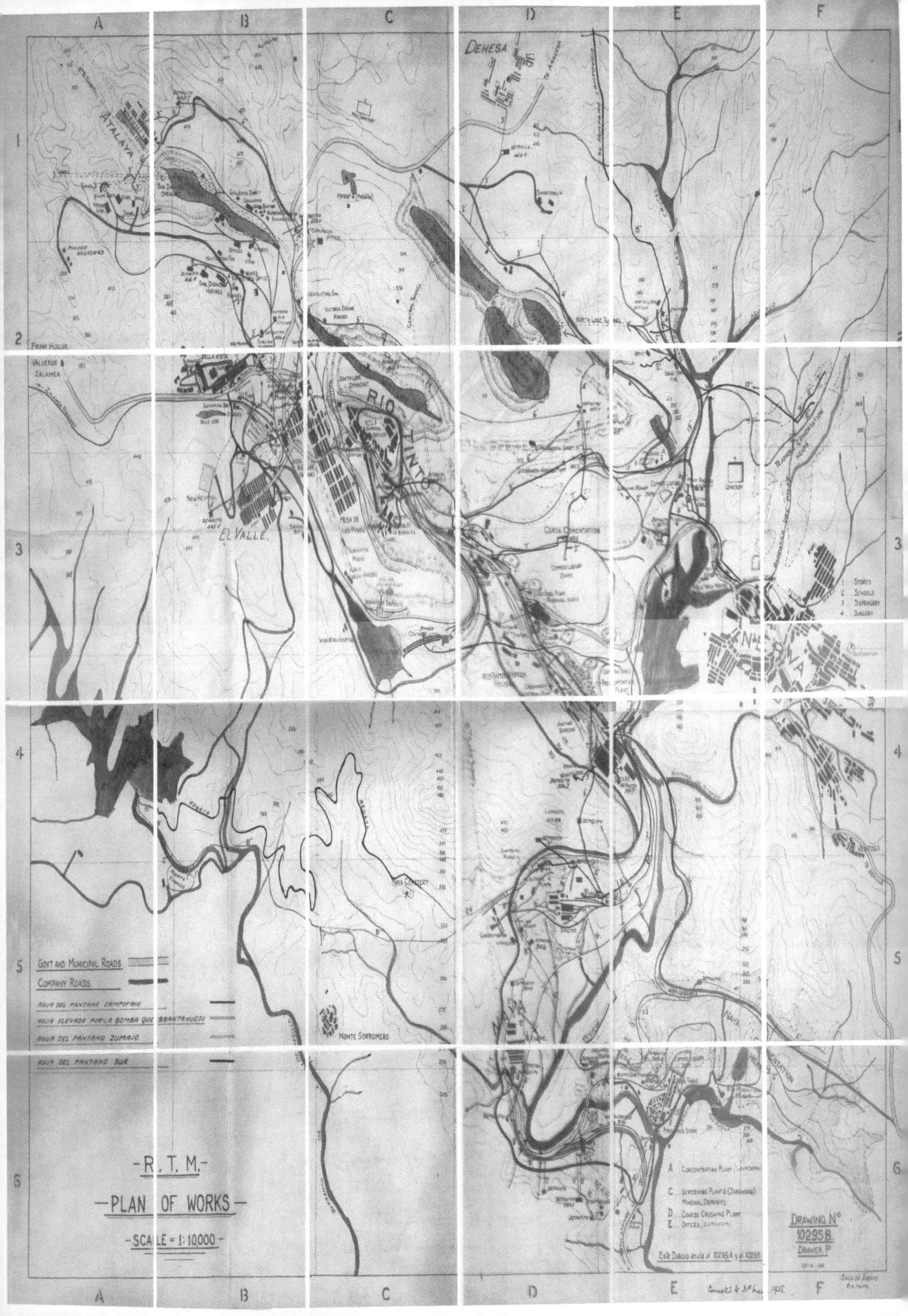

A
B
C
D
E
F
1
2
3
4
5
6
DEHESA
To Aracena
ATALAYA
Powder Magazines
San Dionisio Houses
Offices
Sanatorium
Ventilating Shaft
Victoria Engine Houses
North Lode Tunnel
From Huelva
Valverde
Zalamea
Bella Vista
South Lode Openings
RIO TINTO
New Hospital
EL VALLE
Mesa de los Pinos
Cerda Cementation Plant
Cemetery
1 Stores
2 Schools
3 Dispensary
4 Surgery
NERVA
Ventoso
Nerva Cemetery
Monte Sorromero
Govt and Municipal Roads
Company Roads
Agua del Pantano Campofrio
Agua elevada por la bomba que brantahueso
Agua del Pantano Zumajo
Agua del Pantano Sur
-R. T. M.-
-PLAN OF WORKS-
-SCALE = 1: 10,000-
A... Concentrating Plant (Lavadora)
C... Screening Plant & (Zarandas) Mineral Deposits
D... Coarse Crushing Plant
E... Offices (Zarandas)
Este Dibujo anula al 10295A y al 10295
Drawing No
10295B.
Drawer P.

MIX
Papier aus verantwortungsvollen Quellen
Paper from responsible sources
FSC® C105338

If you have any concerns about our products,
you can contact us on
ProductSafety@springernature.com

In case Publisher is established outside the EU,
the EU authorized representative is:
Springer Nature Customer Service Center GmbH
Europaplatz 3, 69115 Heidelberg, Germany

Printed by Libri Plureos GmbH
in Hamburg, Germany